# Coal Mine Ground Control

# Coal Mine Ground Control

*Editor*

Marcel Talbot

scitus
academics

**Coal Mine Ground Control**
Edited by **Marcel Talbot**

ISBN: 978-1-68117-259-0
Library of Congress Control Number: 2016934802


SCITUS Academics LLC,
www.scitusacademics.com
Box No. 4766, 616 Corporate Way,
Suite 2, Valley Cottage,
NY 10989



Printed in United States of America on Acid Free Paper

# Preface

The goal of coal mining is to obtain coal from the ground. Coal is valued for its energy content, and, since the 1880s, has been widely used to generate electricity. Steel and cement industries use coal as a fuel for extraction of iron from iron ore and for cement production. Coal mining has had many developments over the recent years, from the early days of men tunneling, digging and manually extracting the coal on carts, to large open cut and long wall mines. Mining at this scale requires the use of draglines, trucks, conveyors, jacks and shearers. The most economical method of coal extraction from coal seams depends on the depth and quality of the seams, and the geology and environmental factors. Coal mining processes are differentiated by whether they operate on the surface or underground. Surface mining and deep underground mining are the two basic methods of mining. The choice of mining method depends primarily on depth of burial, density of the overburden and thickness of the coal seam. Seams relatively close to the surface, at depths less than approximately 180 ft (50 m), are usually surface mined. Coal that occurs at depths of 180 to 300 ft (50 to 100 m) is usually deep mined, but in some cases surface mining techniques can be used. Most coal seams are too deep underground for opencast mining and require underground mining, a method that currently accounts for about 60 percent of world coal production. This book entitled Coal Mine Ground Control covers every major topic concerned with ground control in underground coal mines, with emphasis on the application of various ground control theories and techniques. The book reviews current general practices in underground mine layouts, roof control, and common ground control failures, and the discussion of the state-of-the-art of coal mine ground control technology. The book will also prove useful to practical engineers involved in either the design or the day-to-day operations of underground coal mines.

# Table of Contents

# CHAPTER 1

# Optimization Study of Outburst Prevention Measures for Tuzhu Coal Mine Based on Fixed Weight Clustering Analysis

**Wenke Luo[1,2], Shiliang Shi[1,2], Yi Lu[1,2*], Shenghua Zou[1,2], Zaian Chen[3], Liliang Chen[4]**

[1]*Hunan Provincial Key Laboratory of Safe Mining Techniques of Coal Mines, Xiangtan, China*
[2]*School of Energy and Safety Engineering, Hunan University of Science and Technology, Xiangtan, China*
[3]*Jinzhushan Mining Company of Hunan Coal Group, Lenshuijiang, China*
[4]*Baishanping Mining of Hunan Black Gold Era Co., LTD, Leiyang, China*

## ABSTRACT

Affected by many involved factors, different dimensions, data with large difference, incomplete information and so on, the most optimal selection of regional outburst prevention measures for outburst mine has become a complicated system project. The traditional way of outburst prevention measure selection belongs to qualitative method, which may cause high-cost of gas control, huge quantities of drilling work, long construction time and even secondary disaster. To solve the above-mentioned problems, in light of occurrence status of coal seam gas in No. 21 mining area of Jinzhushan Tuzhu Mine, through grey fixed weight clustering theory and a combination method of qualitative and quantitative analysis, the judging model with multi-objective classification for optimization of outburst prevention measures was established. The three weight coefficients of outburst prevention technology scheme are sorted, in order to determine the advantages and disadvantages of each outburst prevention technology scheme under the comprehensive evaluation of multi-target. Finally, the problem of quantitative selection for regional outburst prevention technology scheme is solved under the situation of multi-factor mode and incomplete information, which provides reasonable and effective technical measures for prevention of coal and gas outburst disaster.

## KEYWORDS

Coal-Gas Outburst, Grey Theory, Fixed Weight Clustering Analysis, Regional Outburst Prevention Measures

## 1. INTRODUCTION

China is not only the country with the highest coal production from underground mining in the world, but also the main coal production country with the highest percentage mines with high gas and coal-gas outburst [1] -[3] . With the sustained growth of energy demand, depth of coal mining, coal output, intensive production, mining intensity and gas outburst disaster, it has increased along with the continuous development of underground mining [4] [5] . As a result of the increasing demand for energy especially coal, most mines in China have entered into deep mining with a speed of 20 - 50 m/a [6] [7] . With the increase of mining depth, gas pressure and gas content, coal and gas outburst disasters are on the acceleration. Meanwhile, China has become the country with the most serious disasters of coal-gas outburst in the world. Through decades of theoretical research and engineering practice, the "four-in-one" comprehensive outburst prevention measure has been formed, which greatly enhanced the security of outburst coal seam mining [8] [9] . Regional outburst prevention measures include pre-drainage coal seam sector gas by surface drilling, pre-drainage coal seam sector gas by layer-through boring in floor roadway, pre-drainage coal seam sector gas by drilling along seam, pre-drainage coal roadway strips gas by layer-through boring, pre-drainage region gas when rock cross-cut coal uncovering by layer-through boring, pre-drainage gas in mining region by layer-through boring, pre-drainage gas in mining region by drilling along seam and so on [10] - [15] . On this basis, for the coal seam group mining, a variety of gas pre-drainage measures have evolved such as progressive bands with layer-through boring in floor roadway combined with pre-drainage area gas or strips gas by drilling along seam. During the traditional selection of regional outburst prevention measures, chief engineer of mining company chooses one of the feasible solutions and then approves to implement it. Due to the limitation of personal preference and knowledge level of chief engineers, the final implementation plan is not necessarily the best one. It has been found that there are still mine outbursts after the implementation of outburst prevention measures and there are also other disasters during the implementation process. Therefore, according to the fixed weight method of grey theory, for gas occurrence condition of specific areas, optimizing the preliminary feasible scheme with the multi-objective of the pre-drainage drilling quantities, drilling time, capital investment of gas control, the comprehensive cost of gas treatment and the control ability of secondary disasters will be important for outburst prevention and disaster reduction.

# 2. PROJECT OVERVIEW AND PRELIMINARY PLAN

## 2.1. Overview of No. 21 Mining Area of Tuzhu Mine

According to geological data, the No. 21 mining area is located in the middle of the southeastern wing of Qiaotouhe syncline in Jinzhushan Tuzhu Mine. The mining area, from 26th exploration line in the west to 200 meters east to 29th exploration line in the east, with the strike length of 1350 meters, dip length of 800 meters, covers a total area of 1080 square meters. The upper level of mining is −50 meters and the lower level is −250 meters. The mining area contains seven coal seams named the 1st, 2nd, 3rd, 4th, 5th, 6th and 7th coal seam, respectively. The 5th coal seam is the main mining coal seam, 2nd, 3rd and 4th coal seams could be exploited partly, while 1st, 6th and 7th couldn't be exploited.

## 2.2. Preliminary Plan for Regional Outburst Prevention Measures of No. 21 Minging Area

1) Plan I: Pre-drainage gas in 5th coal seam by lots of crossing holes in floor roadway

That includes laying out one or two rock roadways in depth of 20 - 25 meters in the 5th coal seam floor within the No. 21 mining area, then constructing a drilling site every 50 meters in the rock roadway and lastly drilling holes in grid-style from bottom to top, as shown in Figure 1.

2) Plan II: Pre-drainage gas in the 5th coal seam by layer-through boring bands in floor roadway combined with drilling along seam

It includes laying out floor rock roadway in depth of 20 - 25 meters in the 5th coal seam floor within the No. 21 mining area, then constructing a drilling site in the rock roadway and lastly constructing grid-style crossing hole to coal roadway, striving to eliminate the danger in coal roadway and its surrounding coal in a relatively short time by gas pre-drainage. Thus, it can be seen that gas pre-drainage by stripe drilling hole in floor rock roadway serve for tunneling of coal roadway, as shown in Figure 2.

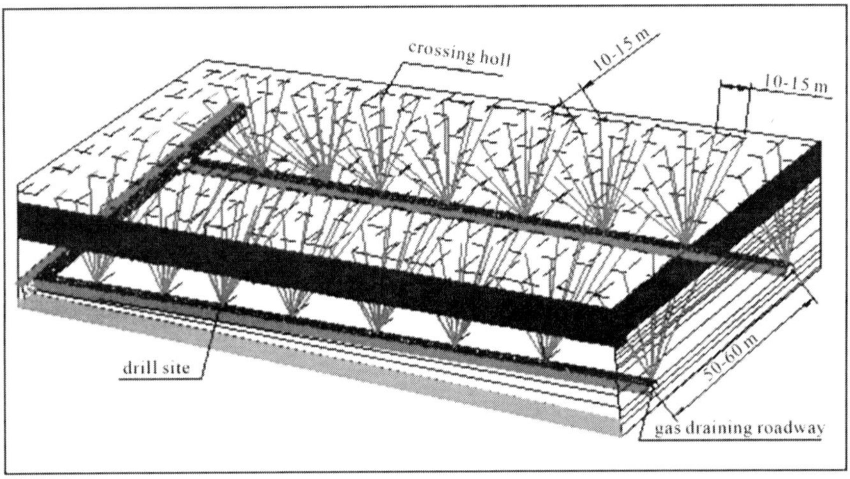


**Figure 1**. Sketch of crossing holes with grid-style layout in floor roadway.

3) Plan III: Progressive shielding gas pre-drainage in coal roadway

That includes constructing crossing hole in floor of 5th coal seam of No. 21 mining area to carry out striped gas pre-drainage for haulage roadway and return airway of 5th seam in the beginning; and then carrying out stripe regional gas pre-drainage for haulage roadway or return airway in 4th coal seam sector after the checksum meets the requirements. After the pressure meets the requirements of Outburst Prevention Regulations, constructing crossing hole in haulage roadway or return airway in 4th coal seam sector to carry out striped gas pre-drainage haulage roadway or return airway in 3th coal seam sector. After the pressure meets the requirements of Outburst Prevention Regulations, constructing crossing hole by the same method to carry out striped gas pre-drainage No. 2 coal seam (see Figure 3).

### 2.3. Calculation of Optimizing Parameters in Schemes

According to the above-mentioned three outburst prevention schemes on the No.21 mining area of Tuzhu mine, the performance computing of the three schemes are carried out from drilling work, construction time, investment of outburst prevention, gas control cost and prevention of secondary disasters, in order to optimize the most real data. According to the geographic features, the drilling length toward trending direction is 1390 meters and the drilling length toward the inclination is 840 meters (see Table 1).

## 3. CLASSIFICATION MODELS OF OPTIMIZED OBJECTS

### 3.1. Mathematical Description

The relative merits of regional outburst prevention solutions can be set as first class (best), second class (better) and third class (bad). Those three classes should be regarded as 3 standard modes, and outburst prevention solutions can be regarded as recognized objects, and their relative ownership can be finally determined by analyzing relative classification corresponding to 3 standard modes. As relative merits of regional outburst prevention solutions affected by several factors, it is a multi-factor recognition problem. Combined with the basic principles and basis of outburst prevention solutions, the property of judgments target for outburst prevention solutions will be grouped into 5 kinds of impact indicators including the total gas drilling work, construction time, investment of outburst prevention, gas control cost and prevention of secondary disasters. The standard modes are viewed as grey $K_n(n=1,2,3)$ . $X_{ij}(i=1,2,5;j=1,2,\cdots,5)$ is set as the sample value of target i with regard to in- dex j. $F_j^k(\cdot)(j=1,2,\cdots,5;k=1,2,3)$ is set as Whitening weight function of decision-making index to k subclass.

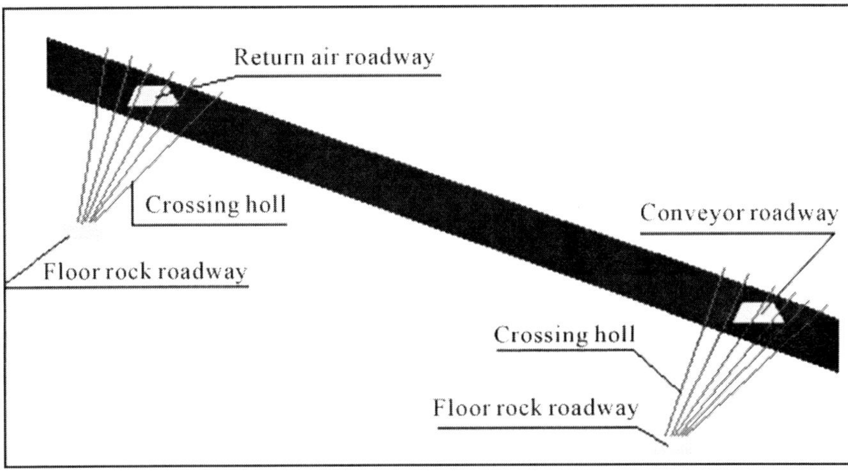


**Figure 2**. Drilling layout for intensive crossing hole and striped pre-drainage in floor rock roadway.

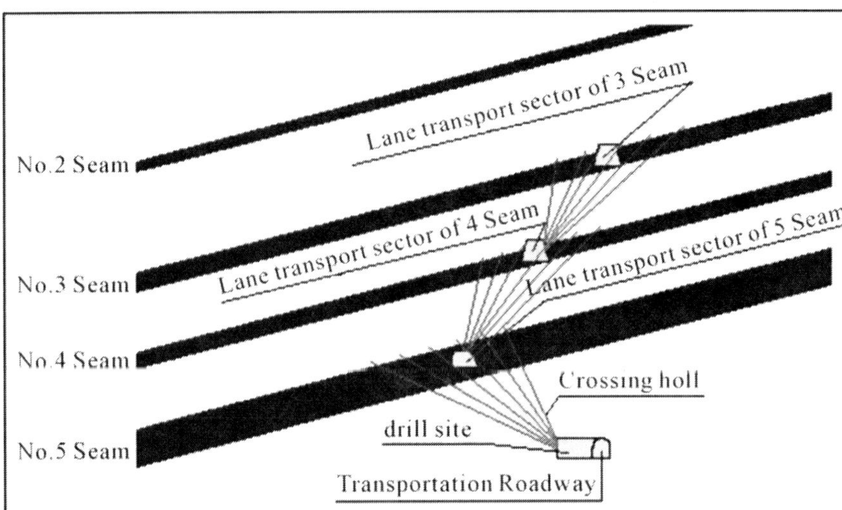


**Figure 3**. Drilling layout for progressive shield-style gas pre-drainage.

**Table 1.** Economic efficiency comparison of preliminary scheme of comprehensive outburst prevention in No. 21 mining area of Jinzhushan Tuzhu mine.

| Outburst prevention schemes | Drilling work ($10^5$ m) | Construction time ($10^3$) | Investment ($10^6$ yuan) | Gas control cost (yuan/t) | Prevention of secondary disasters |
|---|---|---|---|---|---|
| Scheme I | 14.95 | 24.92 | 52.33 | 63.39 | Average (75) |
| Scheme II | 13.54 | 13.54 | 20.31 | 53.29 | Better (80) |
| Scheme III | 15.15 | 15.16 | 22.74 | 82.56 | Good (95) |

Here, function $\eta_j^k\,(j=1,2,3,4,5;\ k=1,2,3)$ of index j to k subclass is nothing with k. Namely, there is always $\eta_j^{k_1}=\eta_j^{k_2}$ to any $k_1,k_2\in\{1,2,3\}$, then the superscript could be omitted. It can be noted as $\eta_j\,(j=1,2,3,4,5)$

and be defined as $\sigma_i^k=\sum_{j=1}^{5} f_j^k\left(x_{ij}\right)\eta_j$, which is the grey fixed weight clustering coefficient of outburst prevention with regard to i to k grey class. If $\sigma_i^{k^*}=\max\limits_{1\le k\le 3}\left\{\sigma_i^k\right\}$, the target i can be judged to belong to grey class $k^*$.

## 3.2. The Division of Grey Classes

The grey classes of the outburst prevention solutions for the No. 21 mining area can be divided into three classes, namely, best, better and bad. According to observations of each index, the standard gray classes of three levels can also be determined as shown in Table 2.

## 3.3. Determination of Whitening Weight Function

The first four of 5 indexes belong to cost variables, and the last index belongs to effective variables. To make calculation simple and the problem clear, the indexes and grey classes should be numbered. The whitening weight function with regard to j index to k subclass $F_j^k(\cdot)(j=1,2,\cdots,5;\ k=1,2,3)$ can be expressed as follows (see Figure 4 and Table 3):

**Table 2.** Division of grey classes for preliminary regional outburst prevention solutions of No. 21 mining area in Jinzhushan Tuzhu mine.

| Grey class | Drilling work ($10^5$ m) | Construction time ($10^3$) | investment (RMB $10^6$) | Gas control cost (Yuan/t) | Prevention of secondary disasters (%) |
|---|---|---|---|---|---|
| Best (I) | 7.5 | 8.9 | 15.9 | 33.2 | 126 |
| Better (II) | 15 | 17.8 | 31.8 | 66.4 | 84 |
| Bad (III) | 22.5 | 26.7 | 47.7 | 99.6 | 42 |

**Table 3.** Whitening weight function $f_j^k(\cdot)$ list with regard to j index to k subclass.

| Evaluation index | $f_j^1(\cdot)$ | $f_j^2(\cdot)$ | $f_j^3(\cdot)$ |
|---|---|---|---|
| Drilling work ($10^5$ m) | $f_1^1(-,-,7.5,15)$ | $f_1^2(7.5,15,-,22.5)$ | $f_1^3(15,22.5,-,-)$ |
| Construction time ($10^3$ days) | $f_2^1(-,-,8.9,17.8)$ | $f_2^2(8.9,17.8,-,26.7)$ | $f_2^3(17.8,26.7,-,-)$ |
| Total investment (RMB $10^6$) | $f_3^1(-,-,15.9,31.8)$ | $f_3^2(15.9,31.8,-,47.7)$ | $f_3^3(31.8,47.7,-,-)$ |
| Control cost (Yuan/t) | $f_4^1(-,-,33.2,66.4)$ | $f_4^2(33.2,66.4,-,99.6)$ | $f_4^3(66.4,99.6,-,-)$ |
| Prevention of secondary disaster (%) | $f_5^1(84,126,-,-)$ | $f_5^2(42,84,-,126)$ | $f_5^3(-,-,42,84)$ |

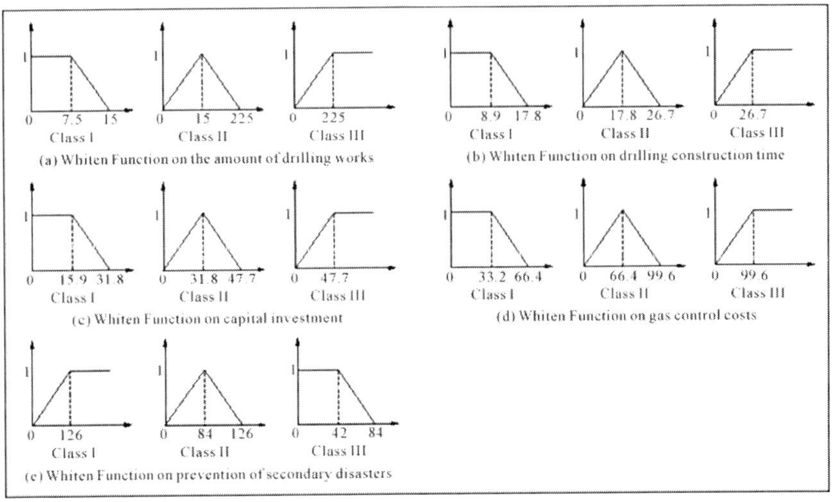


**Figure 4.** Whitening weight function with regard to j index to k subclass.

1) Whitening weight function I

$$f_j^k(x)\begin{cases} 0 & x \prec x_j^k(1) \\ \dfrac{x - x_j^k(1)}{x_j^k(2) - x_j^k(1)} & x \in \left[x_j^k(1), x_j^k(2)\right] \\ 1 & x \geq x_j^k(2) \end{cases} \tag{1}$$

2) Whitening weight function II

$$f_j^k(x)\begin{cases} 0 & x \notin \left[x_j^k(1), x_j^k(4)\right] \\ \dfrac{x - x_j^k(1)}{x_j^k(2) - x_j^k(1)} & x \in \left[x_j^k(1), x_j^k(2)\right] \\ \dfrac{x_j^k(4) - x}{x_j^k(4) - x_j^k(1)} & x \in \left[x_j^k(2), x_j^k(4)\right] \end{cases} \tag{2}$$

3) Whitening weight function III

$$f_j^k(x)\begin{cases} 0 & x \notin \left[0, x_j^k(4)\right] \\ 1 & x \in \left[0, x_j^k(3)\right] \\ \dfrac{x_j^k(4) - x}{x_j^k(4) - x_j^k(3)} & x \in \left[x_j^k(3), x_j^k(4)\right] \end{cases} \tag{3}$$

# 4. CALCULATION OF FIXED WEIGHT CLUSTERING COEFFICIENT

## 4.1. Determination of Clustering Weights

According to the available information of Tuzhu mine, weights of drilling quantities, construction time, investment, gas control cost and prevention of secondary disaster are identified by project team and technical personnel of mine as: $\eta_1 = 0.1512$, $\eta_2 = 0.1512$, $\eta_3 = 0.1441$, $\eta_4 = 0.4094$, $\eta_5 = 0.1441$.

## 4.2. Calculation of Clustering Coefficient

It can be calculated from formula 1: when i = 1 and k = 1, 2, 3,

$$\begin{aligned}\sigma_1^1 &= \sum_{j=1}^{5} f_j^1(x_{ij})\cdot\eta_j = f_1^1(14.95)\times 0.1512 + f_2^1(24.95)\times 0.1512\\ &+ f_3^1(52.33)\times 0.1441 + f_4^1(63.39)\times 0.4094 + f_5^1(75)\times 0.1441\\ &= \frac{14.95-7.5}{7.5}\times 0.1512 + 0 + 0 + \frac{63.39-33.2}{33.2}\times 0.0.4094\\ &= 0.5224.\end{aligned}$$

when i = 1 and k = 2,

$$\begin{aligned}\sigma_1^2 &= \sum_{j=1}^{5} f_j^2(x_{ij})\cdot\eta_j = f_1^2(14.95)\times 0.1512 + f_2^2(24.95)\times 0.1512\\ &+ f_3^2(52.33)\times 0.1441 + f_4^2(63.39)\times 0.4094 + f_5^2(75)\times 0.1441\\ &= \frac{14.95-7.5}{7.5}\times 0.1512 + \frac{24.95-17.8}{17.8}\times 0.1512 + 0 + 1\times 0.4094 + \frac{75-42}{42}\times 0.1441\\ &= 0.7333.\end{aligned}$$

when i = 1 and k =2,

$$\begin{aligned}\sigma_1^3 &= \sum_{j=1}^{5} f_j^3(x_{ij})\cdot\eta_j = f_1^3(14.95)\times 0.1512 + f_2^3(24.95)\times 0.1512\\ &+ f_3^3(52.33)\times 0.1441 + f_4^3(63.39)\times 0.4094 + f_5^3(75)\times 0.1441\\ &= \frac{14.95-7.5}{7.5}\times 0.1512 + 1\times 0.1512 + 1\times 0.1441\\ &+ \frac{63.39-33.2}{33.2}\times 0.4094 + \frac{75-42}{42}\times 0.1441\\ &= 0.9309.\end{aligned}$$

So,

$$\sigma_1 = \left(\sigma_1^1, \sigma_1^2, \sigma_1^3\right) = (0.5224, 0.7333, 0.9039)$$

It also can be calculated that,

$$\sigma_2 = \left(\sigma_2^1, \sigma_2^2, \sigma_2^3\right) = (1.000, 0.6187, 0.6324),$$

$$\sigma_3 = \left(\sigma_3^1, \sigma_3^2, \sigma_3^3\right) = (0.7977, 0.6330, 0.8532).$$

## 4.3. Clustering Analysis

Every preliminary outburst prevention solutions corresponding to k class has a relevant clustering coefficient. In clustering coefficients, the level corresponding to the coefficient with the largest value is the one that clustering samples belong to. According to above results, the grey clustering coefficient matrix can be expressed as follows:

$$\Sigma = \left(\sigma_i^k\right) = \begin{bmatrix} \sigma_1^1 & \sigma_1^2 & \sigma_1^3 \\ \sigma_2^1 & \sigma_2^2 & \sigma_2^3 \\ \sigma_3^1 & \sigma_3^2 & \sigma_3^3 \end{bmatrix} \begin{bmatrix} 0.5224 & 0.7333 & 0.9039 \\ 1.000 & 0.6187 & 0.6324 \\ 0.7977 & 0.6330 & 0.8532 \end{bmatrix}$$

It can be inferred from the above-mentioned grey fixed weight clustering matrix that

$$\max_{1 \le k \le 3}\left\{\sigma_1^k\right\} = \sigma_1^3 = 0.9039,\ \max_{1 \le k \le 3}\left\{\sigma_2^k\right\} = \sigma_2^1 = 1.000,\ \max_{1 \le k \le 3}\left\{\sigma_3^k\right\} = \sigma_3^3 = 0.8532.$$

Judging from the comprehensive

evaluation results of drilling work, construction time, investment, gas control cost and prevention of secondary disaster in the preliminary outburst preventions of No. 21 mining area, the second solution is the best, the first is better, while the third one is relatively poor.

Thus, the second regional outburst prevention solution would be implemented in the Tuzhu Mine, namely, the outburst prevention technology of floor layer-through strip combined with drilling along coal seam mode would be carried out in the No. 21 mining area.

## 5. IMPLEMENTATION EFFECT

To investigate the implementation effect of regional outburst prevention solution in No. 21 mining area of Tuzhu Mine, the $1^{\#}$ and $2^{\#}$ boreholes were drilled at $4^{\#}$ and $6^{\#}$ drilling site in floor roadway to observe the gas pressure relief effect in No. 5 coal seam. The installation of pressure monitoring hole is as shown in Figure 5 in order to detect regional gas pressure relief after the implementation of the scheme II.

The inspection time lasted from March 1 to March 31, 2012. Data was collected once every four days from each pressure gauge. As shown in Figure 6, the gas pressures of $1^{\#}$ and $2^{\#}$ boreholes in $6^{\#}$drilling site were 1.2 MPa in the first observation. After the gas drainage last 18 days, the coal seam gas pressure dropped below 0.74 MPa specified in Regulations on Prevention and Control of Coal and Gas Outburst, while the gas pressures of $1^{\#}$ and $2^{\#}$ boreholes in the $4^{\#}$ drilling site were 1.5 MPa and 1.4 MPa respectively in the first observation, and the gas pressures of $1^{\#}$ and $2^{\#}$ boreholes dropped below the requirement of Regulations on Prevention and Control of Coal and Gas Outburst after 25 days and 28 days respectively.

## 6. CONCLUSIONS

There are many kinds of regional outburst prevention measures. How to choose the correct regional outburst prevention measures is important for the normal replacement of working face, the decreasing the drilling quantities of outburst prevention and saving the cost of outburst prevention. In light of occurrence status of coal seam gas in No. 21 mining area of Jinzhushan Tuzhu mine, through grey fixes weight clustering theory and a combination method of qualitative and quantitative analysis, the judging model with multi-objective classification for optimization of outburst prevention measures is established.

**Figure 5.** Drilling hole installation for relief effect test in the No. 21 mining area.

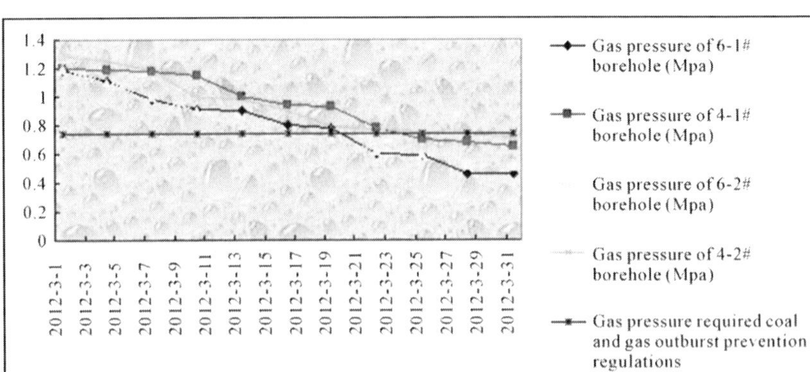


**Figure 6.** Variation curve of boreholes gas pressure in 6# and 4# drilling site.

In particular, the three feasible preliminary regional outburst prevention measures are stated as follows. The first is pre-drainage gas in 5th coal seam by lots of crossing holes in floor roadway; the second is pre-drainage gas in the 5th coal seam by layer-through boring bands in floor roadway combined with drilling along seam and the third is progressive shielding gas pre-drainage in coal roadway. Optimized objects are drilling work, construction time, investment of outburst prevention, gas control cost and prevention of secondary disasters.

The results show that the second measure is the best, the first is better, while the third one is relatively poor. Based on the application effect, the unloading speed of second measure is quick, the unloading time is short. It provides a powerful guarantee for the normal replacement of coal mining working face.

## ACKNOWLEDGEMENTS

This work was supported by the Nation Science Foundation Project (51274100, 51274098), Hunan Province Science and Technology Plan General Projects (2012FJ4268) and Hunan Provincial Education Department General Project (13C307).

## CITE THIS PAPER

WenkeLuo,ShiliangShi,YiLu,ShenghuaZou,ZaianChen,LiliangChen, (2016) Optimization Study of Outburst Prevention Measures for Tuzhu Coal Mine Based on Fixed Weight Clustering Analysis.*Journal of Geoscience and Environment Protection*, **04**, 153-161. doi: 10. 4236/ gep.2016.41016

## REFERENCES


1. Pan, Y. and Li, A. (2013) Rapid Iterative Incremental Model of the Intermittent Chaos of Deep Hole Developing in Coal-Gas Outburst. International Journal of Mining Science & Technology, 23, 287-292.

2. Otuonye, F. and Sheng, J. (1994) A Numerical Simulation of Gas Flow during Coal/Gas Outbursts. Geotechnical & Geological Engineering, 12, 15-34.

3. Cai, F. (2012) Numerical Simulation Research on Coal-Gas Outburst in the Process of Development. Applied Mechanics & Materials, 204-208, 3377-3383.

4. Yang, S., Tang, J., Wen, G., Kang, J. and Liu, C. (2012) Coal and Gas Outburst Disaster Early Warning and Emergency Response Decision Support Technology. Journal of Chongqing University, 35, 121-125.

5. Liu, J., Wang, E., Song, D., Wang, S. and Niu, Y. (2014) Effect of Rock Strength on Failure Mode and Mechanical Behavior of Composite Samples. Arabian Journal of Geosciences, 8, 1-13.

6. Chen, X., Li, W. and Yan, X. (2012) Analysis on Rock Burst Danger When Fully-Mechanized Caving Coal Face Passed Fault with Deep Mining. Safety Science, 50, 645–648.

7. Konicek, P., Soucek, K., Stas, L. and Singh, R. (2013) Long-Hole Destress Blasting for Rock-Burst Control during Deep Underground Coal Mining. International Journal of Rock Mechanics & Mining Sciences, 61, 141-153.

8. Hao, F., Liu, M. and Sun, L. (2011) Study on Comprehensive Technology of Coal and Gas Outburst Prevention in Mining Panels. Procedia Engineering, 26, 699-703.

9. Shen, R., Wang, E., Xu, W.Q., Feng, X. and Hu, S.B. (2011) Outburst Prevention and Control Technology of Thin Coal Zone in Island Coal Face with Potential Outburst. Procedia Engineering, 26, 56-62.

10. Chen, L.W. and Chen, J.C. (2012) Technology of Strip Gas Pre-Drainage with Boreholes along Coal Seam from Front Traverse. Advanced Materials Research, 524-527, 752-757.

11. Jiang, J., Cheng, Y., Zhang, P., Jin, K., Cui, J. and Du, H. (2015) Cbm Drainage Engineering Challenges and the Technology of Mining Protective Coal Seam in the Dalong Mine, Tiefa Basin, China. Journal of Natural Gas Science & Engineering, 24, 412-424.

12. Zou, Q.L., Lin, B.Q., Liu, T., Zhou, Y., Zhang, Z. and Yan, F.Z. (2014) Variation of Methane Adsorption Property of Coal after the Treatment of Hydraulic Slotting and Methane Pre-Drainage: A Case Study. Journal of Natural Gas Science & Engineering, 20, 396-406.

13. Lu, Y. and Qin, B. (2015) Identification and Control of Spontaneous Combustion of Coal Pillars: A Case Study in the Qianyingzi Mine, China. Natural Hazards, 75, 2683-2697.

14. Lu, Y. and Qin, B. (2015) Mechanical Properties of Inorganic Solidified Foam for Mining Rock Fracture Filling. Materials Express, 5, 291-299.

15. Lu, Y., Shen, H., Qin, B., Zhang, L., Ma, H. and Mao, T. (2015) Gas Drainage Radius and Borehole Distance along Seam. Journal of Mining & Safety Engineering, 32, 156-162.

# CHAPTER 2

# Research on and Design of a Self-Propelled Nozzle for the Tree-Type Drilling Technique in Underground Coal Mines

**Yiyu Lu [1,2], Zhe Zhou [1,*], Zhaolong Ge [1,2,*], Xinwei Zhang [1,2] and Qian Li [1,2]**

[1] *State Key Laboratory of Coal Mine Disaster Dynamics and Control, Chongqing University, Chongqing 400044, China*

[2] *National & Local Joint Engineering Laboratory of Gas Drainage in Complex Coal Seam, Chongqing University, Chongqing 400044, China*

## ABSTRACT

Due to the increasing depths of coal mines and the low permeability of some coal seams, conventional methods of gas drainage in underground mines are facing many problems. To improve gas extraction, a new technique using water jets to drill tree-type boreholes in coal seams is proposed. A self-propelled water-jet drilling nozzle was designed to drill these boreholes. The configuration of the self-propelled nozzle was optimized by conducting drilling experiments and self-propelling force measurements. Experimental results show that the optimal self-propelled nozzle has a forward orifice axial angle at 25°, a radial angle at 90°, a center distance of 1.5 mm, and backward pointing orifices with an axial angle of 25°. The self-propelling force generated by the jets of the nozzle with 30 MPa pump pressure can reach 29.8 N, enough to pull the hose and the nozzle forward without any external forces. The nozzle can drill at speeds up to 41.5 m/h with pump pressures at 30 MPa. The radial angles of the forward orifices improve the rock breaking performance of the nozzle and, with the correct angle, the rock breaking area of the orifices overlap to produce a connecting hole. The diameter of boreholes drilled by this nozzle can reach 35.2 mm. The nozzle design can be used as the basis for designing other self-propelled nozzles. The drilling experiments demonstrate the feasibility of using the tree-type drilling technique in underground mines.

## KEYWORDS

Coalbed methane; high-pressure water jet; drilling technique; self-propelled nozzle

## 1. INTRODUCTION

Coalbed methane (CBM) has become a topic of great interest worldwide in recent years. CBM is a kind of unconventional clean energy. The Chinese government attaches great importance to CBM exploitation because CBM provides comprehensive benefits. Its use can increase energy supplies, reduce greenhouse gas emissions, and its extraction can help prevent gas disasters in underground coal mines [1,2,3]. The most common methods of CBM exploitation are surface well drilling and underground gas drainage. The surface well drilling method has been widely used in many developed countries, such as the USA, Canada, and Australia. However, the main method in China is underground gas drainage in order to recover gas and coal simultaneously [4,5,6,7].

The primary technology for CBM drainage in underground mines is to drill gas drainage boreholes into the coal seams [8,9]. These boreholes can be drilled vertically from the floor roadways or horizontally in the coal seams. The horizontal boreholes are usually drilled from roadways developed within the coal seam. This means that the roadway must be tunneled before the coal and gas are extracted, a dangerous operation that, in adverse circumstances, is likely to cause coal and gas outbursts [10,11]. The surface drilling technique, however, also faces some serious problems. These problems include: (1) With increasing mine depth, coal seam permeability decreases and the gas extraction rate becomes very low [12,13]; (2) The volume of coal from which the gas can be extracted by a single borehole is small; (3) A very large number of boreholes are required to ensure that no blank areas (volumes from which the gas cannot be drained) remain in the coal seams; (4) Long range boreholes, which normally go through the floor rock seams, take a very long time to drill; (5) In most cases, coal seam roof and floor rocks are very hard, making drilling difficult, and drilling efficiency is very low; (6) Even after the surface boreholes are completed, extracting the expected quantity of gas takes a long time [14].

To expand the volume of coal from which the gas can be extracted using a single borehole and improve gas drainage efficiency, this paper proposes a new technique for gas drainage by using tree-type boreholes in coal seams. Figure 1 shows a schematic of tree-type drilling. The drilling procedures are: (1) Drilling a main borehole into the coal seam from the floor roadway of an underground coal mine. This is the same as drilling a conventional drainage borehole; (2) Pushing a whipstock into position through the main borehole; (3) Putting a self-propelled nozzle into the orifice of the main borehole; this nozzle moves forward into the whipstock under its own self-propelled force; (4) Drilling a sub-borehole perpendicular to the main borehole with the self-propelled nozzle using a high-pressure water jet;

(5) Changing the direction of the whipstock and repeating step 4; (6) Conducting step 5 multiple times to form a tree-type array of holes.

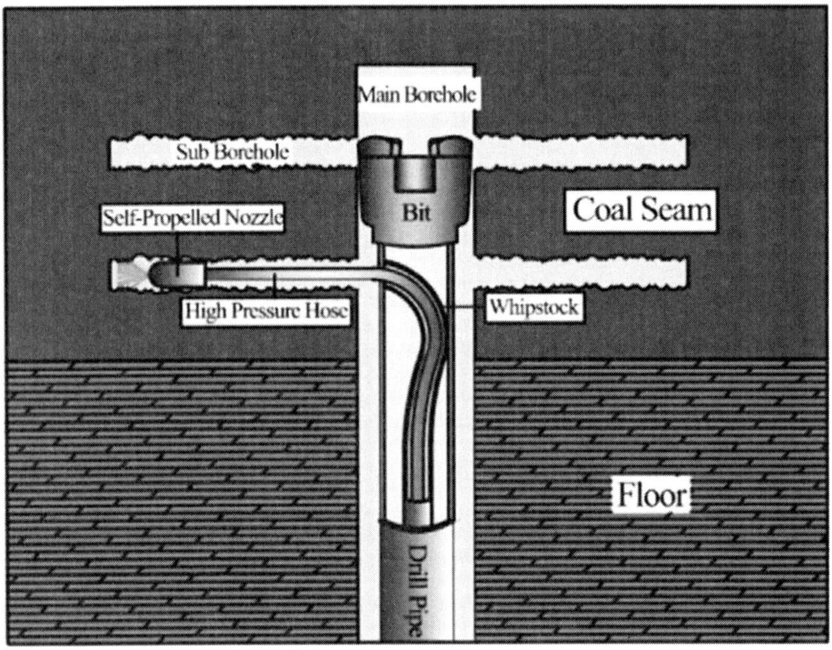


**Figure 1.** Schematic diagram of drilling to make tree-type boreholes in a coal seam.

This technique can create a large number of holes in a coal seam in a relatively short time and the necessary equipment is simpler than that needed for conventional drilling. Using water jets to drill can reduce dust emissions and thus improve working conditions in the mine. The diameter of the sub-borehole can be 30–50 mm and the sub-borehole lengths can reach 10–20 m. The tree-type drilling technique can improve gas drainage efficiency greatly by expanding the volume of coal that can be drained and eliminating the blank areas. However, this technique also faces some challenges. These include design and development of all the equipment needed, choosing the right drilling parameters, and determining the optimal hole arrangements for maximum drainage efficiency.

The equipment needed to drill these tree-type boreholes includes a high-pressure pump, a winch, the whipstock, high-pressure hose, and the self-propelled nozzle. Without doubt, the nozzle is the key to success for tree-type drilling. The nozzle must be able to break the rock (coal) to form a sub-borehole and self-generate enough force to both move forward through the rock and pull the hose behind it.

This paper focuses on the design of the nozzle. The structure of the nozzle was studied in order to obtain a nozzle that could self-propel forcefully. For this purpose experiments to find the optimal nozzle configuration that would break rock efficiently while maintaining its self-propelled ability were conducted.

## 2. PRELIMINARY DESIGN

The self-propelled nozzle is a multi-orifice nozzle with orifices that point both forwards and backwards (Figure 2). The forward pointing orifices direct high-pressure water jets to break the rock (coal) to form the sub-boreholes while the backward pointing orifices eject jets backwards to generate the self-propelling force and to expand the sub-borehole. To be able to manufacture the best nozzle possible, it is necessary to study the rock-breaking properties and the self-propelling ability of the nozzle.

The forward orifices of conventional multi-orifice nozzles are arranged with an axial angle that defines the area of coal that is broken. During drilling, the standoff distance (the distance from forward orifice to the target rock) changes continuously. The axial angle orifice arrangement will cause the area of rock broken by each individual orifice to be separate when the standoff distance is relatively large. This makes breaking the rock in the central zone of target difficult and the result is low drilling efficiency.

This paper proposes adding radial angles to the forward orifices, as shown in Figure 2. Using a radial angle arrangement, the rock breaking area for every orifice will overlap whether the standoff distance is long or short. When the radial angle is 90°, the breaking areas overlap enough to break the central zone of the target rock and form a connecting hole as shown in Figure 3.

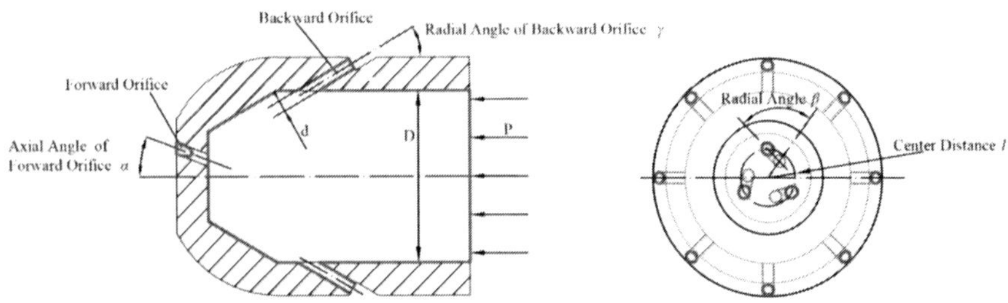


**Figure 2.** Structural diagram of the self-propelled nozzle.

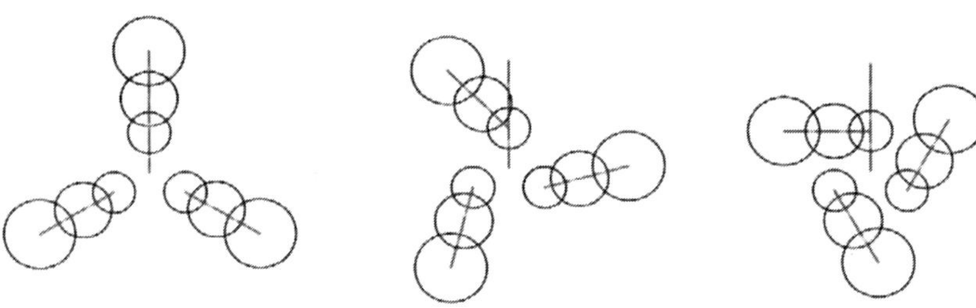

a. Radial Angle 0° b. Radial Angle 45° c. Radial Angle 90°

**Figure 3.** Breaking position changes with standoff distance.

The parameters that must be specified to define the layout of the new multi-orifice nozzle include the diameter and the number of forward and backward orifices, the axial and radial angles of forward orifices, and the radial angle of backward orifices. The diameter and the number of orifices can be obtained by theoretical calculations. The optimal axial and radial angles will be determined by experiments.

Most coal mines in China have gas drainage boreholes under 133 mm in diameter [15]. Because the nozzle and the high-pressure hose have to make turns in the main borehole and the turning radius is limited, the hose should be pliable. The hose selected has an outer diameter of 12.7 mm and of inner diameter is 6.35 mm. Its pressure loss, $\Delta p$, can be calculated by Equation (1) [16,17]:

$$\Delta p = \frac{59.7q^2}{D^5 Re^{0.25}} \tag{1}$$

where $q$ is the volume flow rate in L/min, $D$ is the inner diameter of the high-pressure hose in mm, and $Re$ is the Reynolds number ($Re$ = 11,165 $q/D$).

From Equation (1), it can be seen that the $\Delta p$ increases quickly with an increase in $q$. When $q$ is 40 L/min, $\Delta p$ is 0.63 MPa/m. When $q$ reaches 50 L/min, $\Delta p$ is 0.94 MPa/m. In order to keep the pressure loss reasonably low, 40 L/min was chosen as the design flow rate.

The volume flow rate can be expressed as:

$$q = 60\mu A_t \sqrt{\frac{2p}{\rho}} \tag{2}$$

where $\mu$ is the flow coefficient of the orifice, $A_t$ is the total area of the nozzle orifices in $mm^2$, $p$ is the inlet pressure in MPa, and $\rho$ is the fluid density in $kg/m^3$.

In order to facilitate the machine processing, the orifices in the nozzle are designed to be cylindrical. The flow coefficient for a cylindrical orifice can be valued as 0.82 [18]. The expression of $A_t$ is:

$$A_t = \frac{n\pi d^2}{4} \tag{3}$$

where $n$ is the number of orifices and $d$ is the orifice diameter in mm.

Thus, the Equation (2) can be expressed as:

$d =$ [illegible] (4)

The nozzle has three forward orifices and eight backward facing orifices, so $n$ is 11. Most coal mines in China use model BZW200/31.5 pumps [19] that have a maximum rated capacity of 200 L/min at their maximum rated pressure of 31.5 MPa. Therefore, the inlet pressure for these calculations is taken to be 30 MPa. From Equation (4), the orifice diameter $d$ is 0.62 mm. Due to constraints of actually machining the metal parts, the design value of the orifice diameters is set at 0.6 mm.

As shown in Figure 2, the configuration of the forward orifices is defined by the axial angle, $\alpha$, the radial angle, $\beta$, and the center distance, $l$. These parameters will govern the three-dimensional velocity distribution (axial, radial and tangential) of the jets and determine the rock breaking efficiency of the nozzle. The orientation of the backward orifices is defined only by the axial angle, $\gamma$, and $\gamma$ influences the self-propelling force produced by the jets.

## 3. DRILLING SPEED TESTS

Drilling speed tests were conducted to study the effect of different configurations of the forward orifices on the rock breaking efficiency.

### 3.1. Test Apparatus and Method

Figure 4 is a schematic diagram of the drilling speed test experimental system. Fluid pumped by the high-pressure pump flows through a pressure control valve and then some of fluid flows into the jet generator to form the jets that break the rock. The main components and relevant performance parameters for the experimental system are listed below:

a) The high-pressure pump; rated pressure 32.5 MPa, rated flow rate 200 L/min.

b) The jet generator; main components include the hose connector, the pressure gauge, high-pressure hose, and the nozzle connector.

c) The drilling console mounted on tracks. There is a removable steel bar for fixing the nozzle to the steel pipe connected to the drilling console. When measuring the self-drilling speed, the bar is removed freeing the nozzle to move forward. When conducting the rock breaking tests and studying the pattern of broken holes, the bar is replaced to fix the nozzle to the drilling console so that the two will move together. The rate at which the drilling console can move is controlled by an electric motor and ranges from 0.01 to 100 mm/s.

d) The nozzle. Using the control variable method to design experiments, the influence law of each forward orifice configuration on rock breaking efficiency of the nozzle was studied. Thirteen different self-propelled nozzles were designed and machined for testing. The nozzles, shown in Figure 5, all have the same backward orifices axial angle (25°). The specifications for the forward orifices are shown in Table 1. According to previous research, the optimal axial angle for a nozzle orifice is about 20° for rock breaking [20]. Therefore, the values of axial angle in the nozzles used for this study are around 20° and 5° as a research level considering machining limits. The values of radial angle cover the whole range from 0° to 90°. The values of center distance are limited by the diameter of the nozzle, so the maximum is 1.8 mm (±0.1 mm as a level also due to machining limits).

e) Experimental coal sample. Coals from different mines in different regions can have very different physical properties. In order to test drilling in different kinds of coal and determine if the technique is broadly applicable, large blocks of coal were collected from three coal mines for drilling tests. The mines were the Songzao mine in southwest China, the Pingdingshan mine in central China, and the Tashan mine in northwest China. The coal ranks are from high volatile bituminous to anthracite and their hardnesses range from low to high. The experimental coal block sample from the Pingdingshan mine is shown in Figure 6.

Standard cores were drilled from the coal blocks for mechanical testing. Following International Society for Rock Mechanics (ISRM) standard testing procedures, the 50 mm diameter by 100 mm long cores were dried for 24 h in an oven. The uniaxial compressive strength of the cores were measured by an MTS-815 Universal Testing Machine (MTS Systems Corp., Eden Prairie, MN, USA) [21]. The average strength values for the three samples were 5.1, 10.5 and 15.8 MPa.

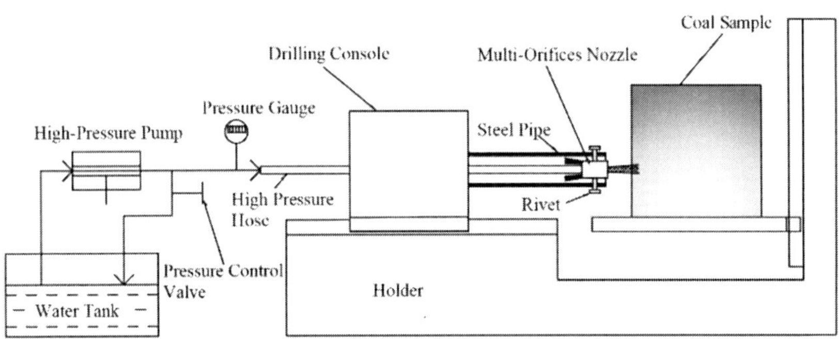


**Figure 4.** Device connection schematic for the self-drilling speed testing system.

All the drilling speed tests were carried out with the same fixed drilling time and fixed jet impact pressure. The drilling test samples are the raw coal blocks from fields, and they are fixed on an open platform during the tests. Drilling time was 90 s and jet impact pressure was 30 MPa. The drilling speed was calculated simply by measuring the depth of the hole drilled and dividing by 90 s.

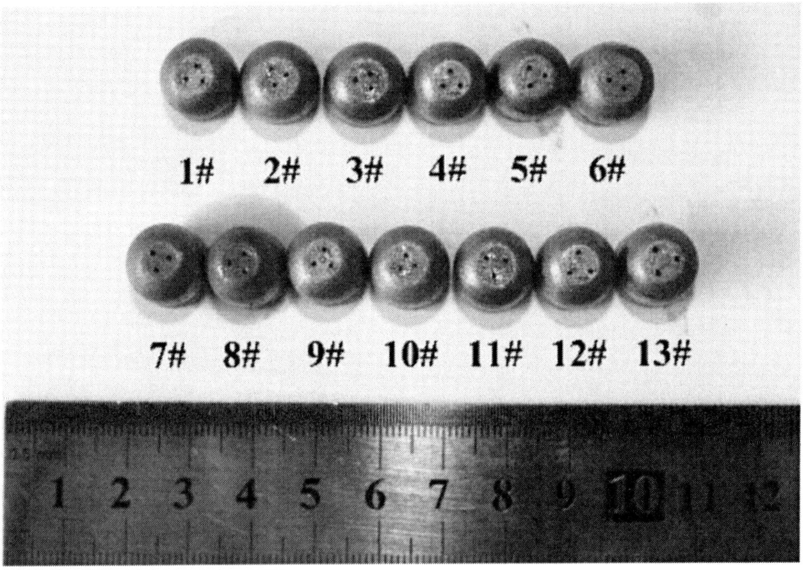


**Figure 5.** Self-propelled nozzles with different forward orifices.

**Table 1.** Axial and radial angles and center distance offsets of the forward orifices for thirteen self-propelled nozzles.

| Number | Forward Orifice Structure Parameters | | |
|---|---|---|---|
| | Axial Angle $\alpha$/° | Radial Angle $\beta$/° | Center Distance $l$/mm |
| $1^{\#}$ | 15 | 0 | 1.5 |
| $2^{\#}$ | 20 | 0 | 1.5 |
| $3^{\#}$ | 25 | 0 | 1.5 |
| $4^{\#}$ | 30 | 0 | 1.5 |
| $5^{\#}$ | 35 | 0 | 1.5 |
| $6^{\#}$ | 25 | 25 | 1.5 |
| $7^{\#}$ | 25 | 45 | 1.5 |
| $8^{\#}$ | 25 | 65 | 1.5 |
| $9^{\#}$ | 25 | 90 | 1.5 |
| $10^{\#}$ | 25 | 90 | 1.4 |
| $11^{\#}$ | 25 | 90 | 1.6 |
| $12^{\#}$ | 25 | 90 | 1.7 |
| $13^{\#}$ | 25 | 90 | 1.8 |

Figure 6. Coal sample from the Pingdingshan mine used for drilling tests.

## 3.2. Experimental Results and Discussion

### 3.2.1. The Effect of the Forward Orifice Axial Angle on Self-Drilling Speeds

The axial angle of the forward orifice determines the jet incidence angle, that is, the angle between the jet axis and the plane normal to the rock surface. When jet fluid hits the rock from different directions, the stress distribution within the rock changes; the axial angle of the forward orifice is an important factor for drilling.

As shown in Figure 7, the range of axial angles is 15°–35°. Under the same experimental conditions, the self-drilling speed first increases and then decreases with the increase of axial angle. Drilling speed is highest when the axial angle of the forward orifice is 25°. When the axial angle is less the 25°, the backflow caused by the forward jets on the rock surface is strong and this hinders the follow-up jets and causes the self-drilling speed to decreasing. When the axial angle is around 25°–35°, the jet impact pressure on the sample decreases gradually with the increase axial angle so the self-drilling speed also gradually decreases. Using the self-drilling speed as the evaluation criterion, the optimal axial angle for forward orifice as determined by these experiments is 25°.

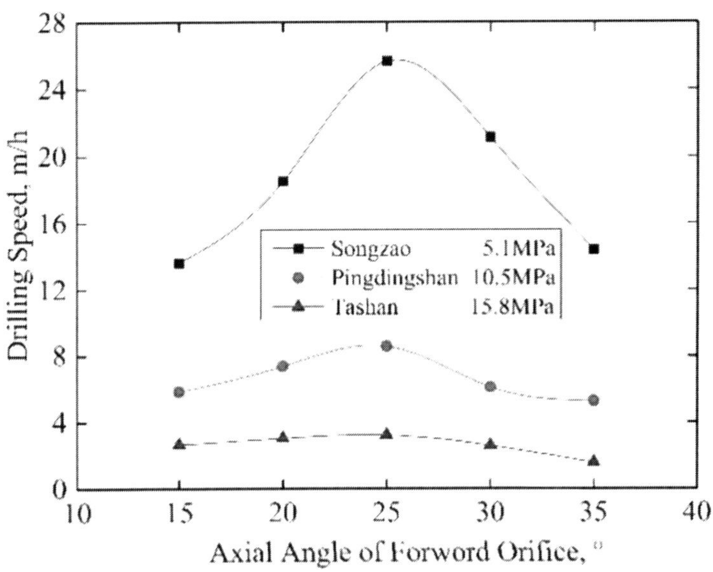


**Figure 7.** Graph showing the relationship between self-drilling speed and forward orifice axial angle. Strength of coal samples: 5.1, 10.5 and 15.8 MPa.

### 3.2.2. The Effect of the Forward Orifice Radial Angle on Self-Drilling Speeds

During self-drilling, the radial angle of forward orifice determines the direction of the jets and that determines the pattern of the holes broken. The pattern that will make the holes overlap more will achieve higher drilling speeds. As shown in Figure 8, the drilling speed increases with the increase of radial angle.

Rock breaking tests were conducted to study the patterns of broken holes using different nozzles with different radial angles. Sandstone was chosen for the target because it is easier to observe the pattern of broken holes in sandstone than in coal. The uniaxial compressive strength of the sandstone sample was 27.4 MPa. The initial standoff distance was 5 mm and the drilling console + nozzle was advanced at 0.1 mm/s. The hole patterns created by four different nozzles are shown in Figure 9.

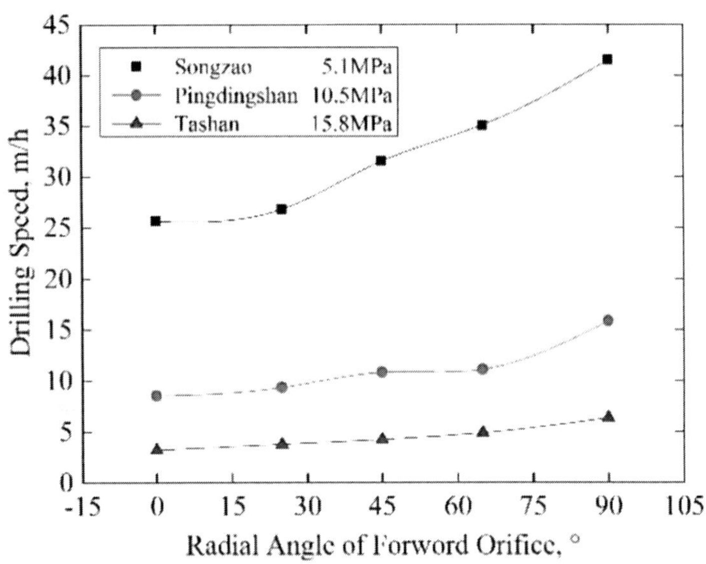


**Figure 8.** Graph showing the relationship between self-drilling speed and forward orifice radial angle. Strength of coal samples: 5.1, 10.5 and 15.8 MPa.

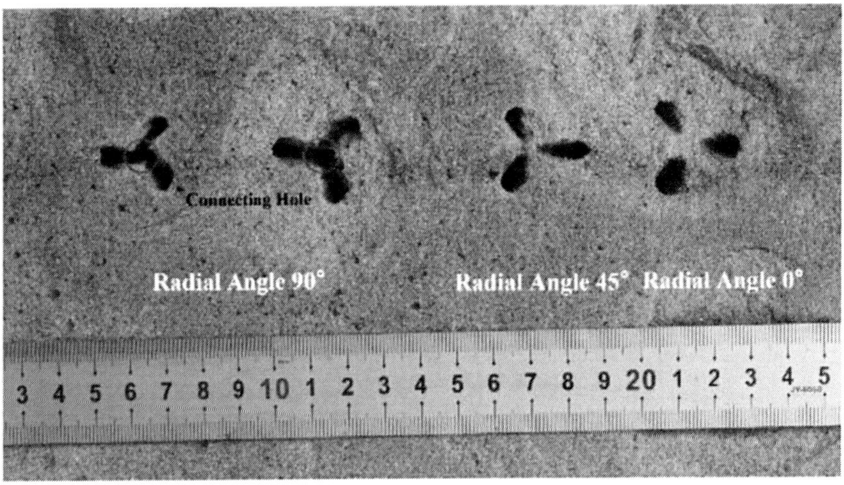


**Figure 9.** Pattern of holes broken in sandstone with different forward orifice radial angles. Note that the nozzles with the 90° radial angle orifices broke out the center zone of the target (red circles) whereas the 45° and 0° radial angle nozzles did not.

When the radial angle of the forward orifices is 90°, the areas of rock broken by the orifices overlap and form a connecting hole. This pattern effectively combines the jet impact of the three forward orifices to break the central zone and is the reason that this 90° radial angle arrangement results in the highest self-drilling speed.

### 3.2.3. The Effect of Forward Orifice Center Distance on Self-Drilling Speeds

The center distance of the forward orifices determines how much the jets diverge and also influences how much the jets and the holes they produce overlap.

As listed in Table 1, the center distances for the nozzle forward orifices range from 1.4 to 1.8 mm. Figure 10 shows self-drilling speed plotted against the center distance for five tests with nozzles all having forward orifice radial angles of 90°. Under the same experimental conditions, the self-drilling speed first increases and then decreases with the increase of center distance. When the center distance is 1.5 mm, the self-drilling speed is at its maximum. When the center distance is 1.4 mm, the decrease in drilling speed is because the arrangement of the three orifices is so centralized that the total area of rock broken is too small. The nozzle does not fit through the hole even though holes broken by the jets overlap and break out the center zone. When the center distance is more than 1.5 mm, the overlap area of the jets decreases as the center distance increases, hindering the self-drilling. For these experiments, the optimal center distance for the forward orifice is 1.5 mm.

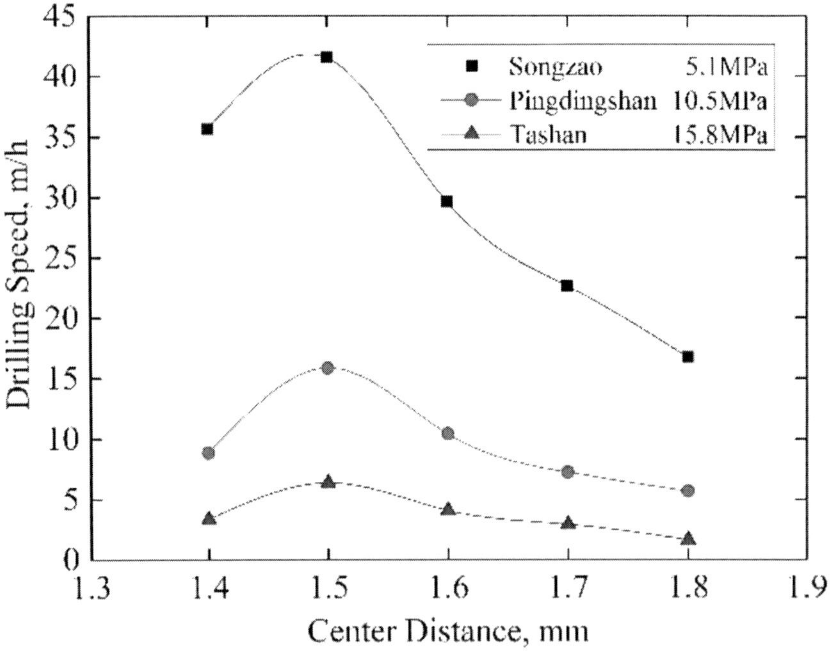


**Figure 10.** Graph showing the relationship between self-drilling speed and forward orifice center distance. Strength of coal samples: 5.1, 10.5 and 15.8 MPa.

# 4. SELF-PROPELLING FORCE MEASUREMENT

Tests were also conducted to measure the self-propelling force that the nozzles could generate. The amount of force that a nozzle can produce is a function of fluid pressure and the nozzles backward orifices axial angle, so total self-propelling force tests were conducted using different backward orifice configurations at different pressures.

## 4.1. Test Apparatus and Method

These experiments used an accurate measuring system for small forces. A schematic of the equipment is shown in Figure 11. The system includes:

a) The high-pressure pump and jet generator. The performance parameters are the same as those described in Section 3.1.

b) A through-hole load cell. A load cell, model number LC8200-625-50 (from Omega Engineering Company, Stamford, CT, USA) was used to measure the forces. The load cell has a ~5 cm (2-inch) outer diameter and ~16 mm (0.625-inch) inner diameter and can measure forces in the 0–222 N range accurately.

c) A glass pipe. The glass pipe was used to simulate a gas drainage borehole in a coal seam. The pipe's inner diameter was 50 mm.

d) Self-propelled nozzles. To analyze the total self-propelling force of the nozzle, five self-propelled nozzles with different backward orifice structures were machined. The nozzles are shown in Figure 12. These five nozzles all have the same forward orifice configuration (axial angle 25°; radial angle 90°; center distance 1.5 mm). The specifications for the backward orifices on the nozzles are shown in Table 2.

Total self-propelling force measurement was carried out at three different pump pressures, 10, 20 and 30 MPa.

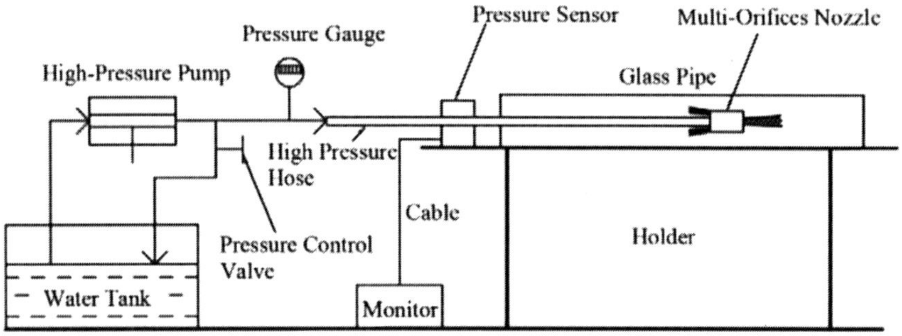


**Figure 11.** Device connection schematic for a self-propelling force measurement system.

**Figure 12.** Self-propelled nozzles with different backward orifices axial angles.

**Table 2.** Backward orifice axial angles for five self-propelled nozzles used for self-propelling force measurements.

| Number | Backward Orifice Axial Angle/° |
|---|---|
| 1# | 10 |
| 2# | 15 |
| 3# | 20 |
| 4# | 25 |
| 5# | 30 |

## 4.2. Experimental Results and Discussion

### 4.2.1. The Effect of Backward Orifice Axial Angle on Total Self-Propelling Force

Figure 13 shows the relationship between the self-propelling force and the axial angle of the backward orifices. When the axial angle is under 25°, the self-propelling force is large. For our experiment, it is 29.82 N for a 25° axial angle with 30 MPa pump pressure. The self-propelling force is acting against the frictional resistance drilling in coal seams and the resistance produced in the whipstock. The laboratory measured friction coefficient between the hose and the hole wall in coal seams is 0.25–0.35. The hose used in the tests weighs 0.2 kg per meter, so the force exerted by gravity is about 20 N when the drilling length is 10 m. The calculated value of frictional resistance is about 7 N. The resistance produced in the whipstock can be measured by the load cell and the value is approximately 10 N. The total force retarding the nozzle advance is 17 N and smaller than the nozzle propelling force. Therefore, the self-propelling force is enough to pull the hose forward without any external forces being provided. When the axial angle is more than 25°, the self-propelling force decreases significantly. When the axial angle of the backward orifice increases, the component of jet velocity parallel to the axis of the nozzle decreases and the self-propelling force decreases as well.

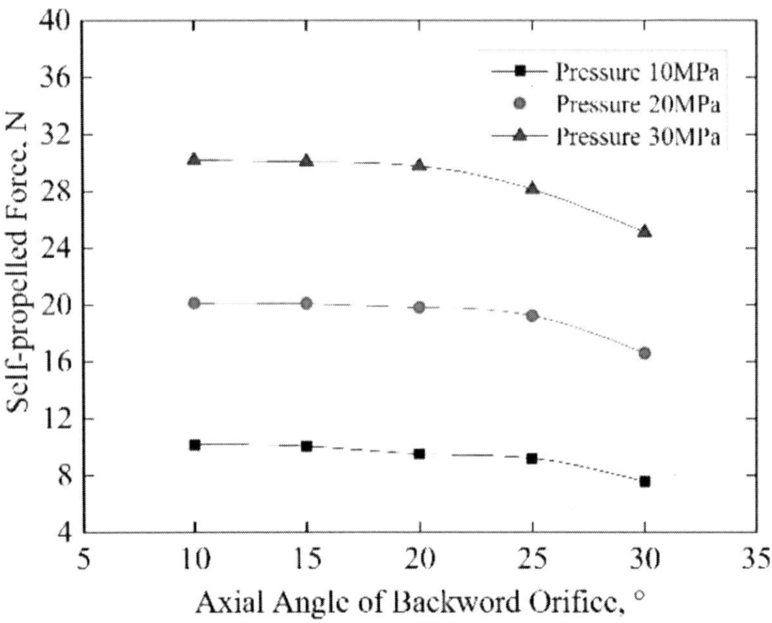


**Figure 13.** Graph showing the relationship between the self-propelling force and the axial angle of the backward orifices.

### 4.2.2. The Effect of the Backward Orifice Axial Angle on the Diameter of the Self-Drilled Borehole

Self-drilling tests using these five nozzles were also conducted. The relationship between the diameter of the self-drilled borehole and the axial angle of backward orifices is shown in Figure 14. Borehole diameter increases as the axial angle of the backward orifices increases. Taking self-propelling force and self-drilling borehole diameter as evaluation criteria, the optimal axial angle for backward orifices is 25°. With the orifices at 25°, the self-propelling force is close to its maximum and the diameter of the borehole is sufficiently large to meet the requirements of a working mine.

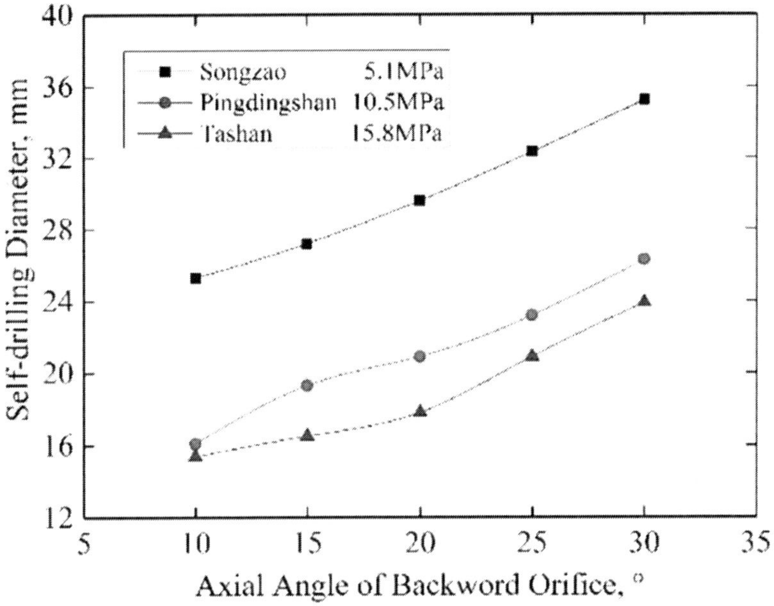


**Figure 14.** Graph showing the relationship between borehole diameter and backward orifice axial angle. Strength of coal samples: 5.1, 10.5 and 15.8 MPa.

## 5. CONCLUSIONS

A new technique using water jets to drill tree-type boreholes in coal seams is proposed to improve gas drainage in underground coal mines. The tree-type drilling technique can be used in coal mines in any country that use boreholes drilled from the floor roadways to drain gas, especially for coal seams of low permeability. The method of drilling sub-boreholes through one main borehole can create boreholes in coal seams to improve gas extraction efficiency in a relatively short time. Furthermore, using the self-propelled water jet nozzle makes the drilling simpler than using conventional drills.

As the most important component, a new self-propelled nozzle was designed and optimized based on the physical conditions in underground mines. The nozzle has three forward orifices and eight backward orifices. The forward orifices can break up the rock to drill sub-boreholes while the backward pointing orifices expel jets to generate the self-propelling force and move the nozzle forward without any external force being applied. The configuration of the nozzle was optimized through experiments. The optimal structure is that the forward orifice axial angle should be 25°, the radial angle 90°, the center distance should be 1.5 mm, and the backward orifice axial angle should be 25°. Unlike traditional multi-orifices nozzles, adding radial angles to the forward orifices improves rock breaking efficiency. When the radial angle of the forward orifices is 90°, the rock breaking area of the orifices overlap and the three jets form holes that connect across the face of the borehole. The diameter of the self-drilled borehole reaches 23.9–35.2 mm, large enough for the nozzle pass though. The self-propelling force generated by the jets of the nozzle under a pump pressure 30 MPa can reach 29.8 N, enough to pull both the high-pressure hose and the nozzle forward. With the pump pressure at 30 MPa, the nozzle can drill at speeds of 6.38–41.5 m/h. The successful drilling of coal samples in the laboratory shows that this tree-type drilling technique is feasible.

## ACKNOWLEDGMENTS

This study was financially supported by the National Natural Science Foundation of China (NSFC) under Grant No. 51374258 and No. 51504046, Program for Changjiang Scholars and Innovative Research Team in University of China under Grant No. IRT13043.

## AUTHOR CONTRIBUTIONS

Yiyu Lu, Zhe Zhou and Zhaolong Ge all contributed to developing the mathematical model, designing the experiments, and writing the paper; Xinwei Zhang and Qian Li performed experiments.

## REFERENCES


1. Dai, J.X.; Ni, Y.Y.; Wu, X.Q. Tight gas in China and its significance in exploration and exploitation. *Pet. Explor. Dev.* **2012**, *39*, 277–284.
2. Shen, C.M.; Lin, B.Q.; Sun, C.; Zhang, Q.Z.; Li, Q.Z. Analysis of the stress–permeability coupling property in water jet slotting coal and its impact on methane drainage. *J. Pet. Sci. Eng.* **2015**, *126*, 231–241.
3. Hu, G.Z.; Xu, J.L.; Zhang, F.X.; Zhao, C.C.; Qin, W.; Zhu, Y.R. Coal and coalbed methane co-extraction technology based on the ground movement in the Yangquan coalfield, China. *Energies* **2015**, *8*, 6881–6897. [Google Scholar] [CrossRef]
4. Sachsenhofer, R.F.; Privalov, V.A.; Panova, E.A. Basin evolution and coal geology of the Donets Basin (Ukraine, Russia): An overview. *Int. J. Coal Geol.* **2012**, *89*, 26–40.
5. Clarkson, C.R. Production data analysis of unconventional gas wells: Review of theory and best practices. *Int. J. Coal Geol.* **2013**, *109–110*, 101–146.
6. Zou, Q.L.; Lin, B.Q.; Liu, T.; Zhou, Y.; Zhang, Z. Variation of methane adsorption property of coal after the treatment of hydraulic slotting and methane pre-drainage: A case study. *J. Nat. Gas. Sci. Eng.* **2014**, *20*, 396–406.

7. Zhang, Y.B.; Gong, B.; Li, J.C.; Li, H.Y. Discrete fracture modeling of 3D heterogeneous enhanced coalbed methane recovery with prismatic meshing. *Energies* **2015**, *8*, 6153–6176.

8. Noack, K. Control of gas emissions in underground coal mines. *Int. J. Coal Geol.* **1998**, *35*, 57–82.

9. Karacan, C.Ö.; Diamond, W.P.; Schatzel, S.J. Numerical analysis of the influence of in-seam horizontal methane drainage boreholes on longwall face emission rates. *Int. J. Coal Geol.* **2007**, *72*, 15–32.

10. Xu, T.; Tang, C.A.; Yang, T.H.; Zhu, W.C.; Liu, J. Numerical investigation of coal and gas outbursts in underground collieries. *Int. J. Rock Mech. Min.* **2006**, *43*, 905–919.

11. Zhu, W.C.; Wei, C.H.; Li, S.; Wei, J.; Zhang, M.S. Numerical modeling on destress blasting in coal seam for enhancing gas drainage. *Int. J. Rock Mech. Min.* **2013**, *59*, 179–190.

12. Lu, T.K.; Yu, H.; Zhou, T.Y.; Mao, J.S.; Guo, B.H. Improvement of methane drainage in high gassy coal seam using waterjet technique. *Int. J. Coal Geol.* **2009**, *79*, 40–48.

13. Lu, Y.Y.; Liu, Y.; Li, X.H.; Kang, Y. A new method of drilling long boreholes in low permeability coal by improving its permeability. *Int. J. Coal Geol.* **2010**, *84*, 94–102.

14. Karacan, C.Ö. Evaluation of the relative importance of coalbed reservoir parameters for prediction of methane inflow rates during mining of longwall development entries. *Comput. Geosci.* **2008**, *34*, 1093–1114.

15. Lu, T.K.; Wang, Z.F.; Yang, H.M.; Yuan, P.J.; Han, Y.B. Improvement of coal seam gas drainage by under-panel cross-strata stimulation using highly pressurized gas. *Int. J. Rock Mech. Min.* **2015**, *77*, 300–312.

16. Labus, T.J. *Fluid Jet Technology: Fundamentals and Application*; Water Jet Technology Association: St. Louis, MO, USA, 1995.

17. Ma, D.J.; Li, G.S.; Huang, Z.W.; Niu, J.L.; Hou, C.; Liu, M.J.; Li, J.B. A model of calculating the circulating pressure loss in coiled tubing ultra-short radius radial drilling. *Pet. Explor. Dev.* **2012**, *39*, 528–533.

18. Frank, M.W. *Fluid Mechanics*, 7th ed.; McGraw-Hill: New York, NY, USA, 2011. Lu, Y.Y.; Cheng, L.; Ge, Z.L.; Xia, B.W.; Li, Q.; Chen, J.F. Analysis on the initial cracking parameters of cross-measure hydraulic fracture in underground coal mines. *Energies* **2015**, *8*, 6977–6994.

19. Shen, Z.H. *Water Jet Theory and Technology*; Petroleum University Publishing House: Dongying, Shandong, China, 1998. (In Chinese)

20. Ma, D.; Miao, X.X.; Chen, Z.Q.; Mao, X.B. Experimental investigation of seepage properties of fractured rocks under different confining pressures. *Rock. Mech. Rock. Eng.* **2013**, *46*, 1135–1144.

# CHAPTER 3

# The Video Collaborative Localization of a Miner's Lamp Based on Wireless Multimedia Sensor Networks for Underground Coal Mines

**Kaiming You, Wei Yang * and Ruisong Han**

*School of Electronic and Information Engineering, Beijing Jiaotong University, Beijing 100044, China*

## ABSTRACT

Based on wireless multimedia sensor networks (WMSNs) deployed in an underground coal mine, a miner's lamp video collaborative localization algorithm was proposed to locate miners in the scene of insufficient illumination and bifurcated structures of underground tunnels. In bifurcation area, several camera nodes are deployed along the longitudinal direction of tunnels, forming a collaborative cluster in wireless way to monitor and locate miners in underground tunnels. Cap-lamps are regarded as the feature of miners in the scene of insufficient illumination of underground tunnels, which means that miners can be identified by detecting their cap-lamps. A miner's lamp will project mapping points on the imaging plane of collaborative cameras and the coordinates of mapping points are calculated by collaborative cameras. Then, multiple straight lines between the positions of collaborative cameras and their corresponding mapping points are established. To find the three-dimension (3D) coordinate location of the miner's lamp a least square method is proposed to get the optimal intersection of the multiple straight lines. Tests were carried out both in a corridor and a realistic scenario of underground tunnel, which show that the proposed miner's lamp video collaborative localization algorithm has good effectiveness, robustness and localization accuracy in real world conditions of underground tunnels.

## KEYWORDS

Wireless multimedia sensor networks (WMSNs); video collaborative localization; underground coal mine; miner's lamp.

## INTRODUCTION

In order to strengthen production safety, the underground coal mine has an urgent demand for wireless multimedia communication services [1,2,3,4,5]. Wireless multimedia sensor networks (WMSNs) are a new kind of sensor networks that introduce multimedia applications, such as images, and audio and video, into wireless sensor networks (WSNs). It is expected that the multimedia monitoring ability will be improved significantly by applying WMSNs in underground coal mine because of the excellent multimedia perception ability, fast and convenient wireless access and flexible topology characteristics of WMSNs [6,7,8,9,10]. In this article, WMSNs in an underground coal mine are structured and a video collaborative localization algorithm for miner is proposed based on the WMSNs to improve the safety of miners.

In WMSNs, video localization of targets is divided into localization with a single camera [11,12] and localization with collaborative cameras [13,14,15,16,17,18,19,20]. It is usually required to obtain the height of target in advance [11] or to extract target perfectly [12] for localization with single camera. Since the altitude localization and plane localization of miner's lamp need to be detected simultaneously, the localization with collaborative cameras are applied in this paper. Video collaborative localization of targets is achieved through fusing the video information collected and processed by multiple cameras. For example, reference [13] proposed a visual collaborative localization algorithm for the passive target positioning problem. According to machine vision theory, the target was projected into several straight lines on the ground plane. The optimum intersection of straight lines, which could be obtained by using Hough transform, was the position estimation of the target. However, because of narrow structures of underground tunnels, cameras can only be deployed along the longitudinal direction of tunnels. Thus, the overlap of projection straight lines may result in location failure with this method.

References [14,15] proposed a method to utilize the active cameras to localize the target collaboratively. The perspective projection model of target was established by utilizing the Gaussian error function. Cameras identify the target location by recognizing the target body and calculating the position and the size of target body based on the perspective projection model. Moreover, the problem of active cameras selection was tackled by balancing the tradeoff between the accuracy of target localization and the energy consumption in camera sensor networks.

Reference [16] presented a technique of multiple feature points to compute the target location in the wireless visual sensor networks. The target edge was determined by Canny detector and the position of target was calculated by fusing the edge information of the target. However, since underground coal mine has feature of insufficient illumination and there may be a lot of coal dusts attaching on miners' uniforms [21], miner's body and the mine's background are very similar, which makes it difficult to distinguish miner's profile from the mine's background.

Therefore, considering insufficient illumination and narrow underground tunnels with bifurcated structures, a miner's lamp video collaborative localization algorithm based on WMSNs was proposed. Since miners must carry cap-lamps to offer auxiliary lighting in underground coal mine which provide a sharp contrast to the dim background, in the algorithm, it was proposed to locate a miner by detecting the cap-lamp of the miner. The miner's lamp project mapping points on the imaging plane of cameras and the coordinates of mapping points are calculated by cameras. Then, multiple straight lines between the positions of cameras and their corresponding mapping points are established. Finally, a miner can be located by finding the optimal intersection of the multiple straight lines. To verify the effectiveness of the proposed miner's lamp video collaborative localization algorithm, tests were carried out both in a corridor and a realistic scenario of underground tunnel. The experimental results show that good localization performance is achieved with the proposed algorithm.

The rest of this article is organized as follows. Section 2 introduces the architecture of WMSNs and location scene for underground coal mine. The algorithm of video collaborative localization of the miner's lamp is discussed in Section 3 and Section 4 presents the experiments results. Section 5 concludes the article.

## 2. ARCHITECTURE OF WMSNS AND LOCATION SCENE FOR UNDERGROUND COAL MINE

Figure 1 shows the architecture of WMSNs in underground coal mine. As shown in Figure 1, the system consists of three layers, *i.e.*, wireless multimedia based networks, wired backbone monitoring bus as well as Ground Monitoring and Dispatching Center (GMDC). The wireless multimedia based networks are constituted by cameras, mobile phones and scalar sensors, *etc.* They are deployed in underground coal mine to collect multimedia information. The multimedia information collected is processed by cluster head nodes and sent to the sink nodes in multi-hop relaying way. Sink nodes which connect with wired backbone monitoring bus transmit multimedia information collected to GMDC via wired backbone monitoring bus. In this way, the GMDC can monitor the multimedia information of underground coal mine in real time.

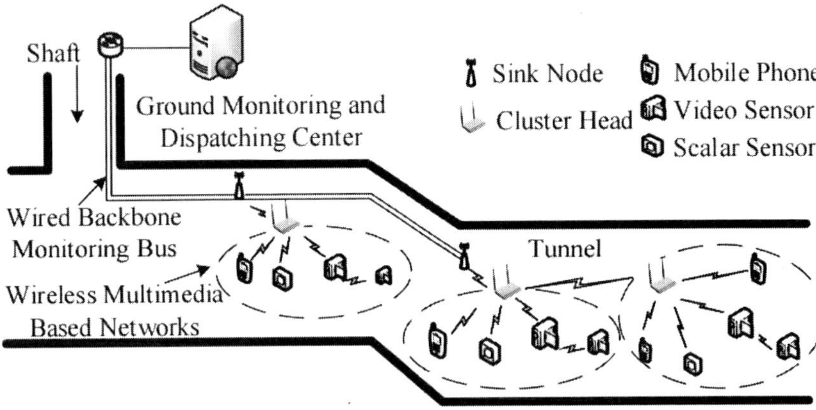


**Figure 1.** Architecture of wireless multimedia sensor networks (WMSNs) in underground coal mine.

Figure 2 shows the scene of miner localization in underground coal mine with bifurcated structures. With the purpose of describing the positions of cameras and miners, a unified World Coordinate System (WCS)$o_w x_w y_w z_w$ is established along underground tunnels with the transverse direction of the Tunnel 1 as $x_w$-axis, the longitudinal direction as $y_w$-axis and the altitude direction as $z_w$-axis, respectively. The tunnel entrance is selected as the origin of the WCS, so that any point in underground tunnels can be located according to its position along the tunnels. It notes that any underground tunnels can be described with the WCS.

In order to locate miners, the WMSNs cameras deployed in underground tunnels should be aware of positions of themselves as well as positions of adjacent cameras which are referred to the geometrical parameters in this paper. In fact, several algorithms of localization and calibration can be applied to obtain the geometrical parameters automatically [22,23,24,25,26]. However, in our experiments, cameras are deployed manually and their geometrical parameters are measured also manually.

The gray part in Figure 2 represents underground tunnels with bifurcated structures. Tunnel 1 and Branch 1, Branch 2 constitute the bifurcation areas of Forks 1 and 2. $C_i$ in Figure 2 represents camera nodes which have the same image processing, digital processing, wireless transmission and reception properties. Cameras $C_1$, $C_2$ and $C_3$ are deployed along the longitudinal direction of tunnels, facing Fork 1 and monitor Fork 1 together. Similarly, cameras $C_4$, $C_5$ and $C_6$ are deployed along the longitudinal direction of tunnels, facing Fork 2 and monitor Fork 2 together. Cameras $C_3$ and $C_4$ are deployed along the longitudinal direction of Tunnel 1, monitoring the connected region of Fork 1 and Fork 2 together. The green shadow part inFigure 2 represents the areas that cameras monitor. The area that one camera can monitor is approximately a pyramid shape.

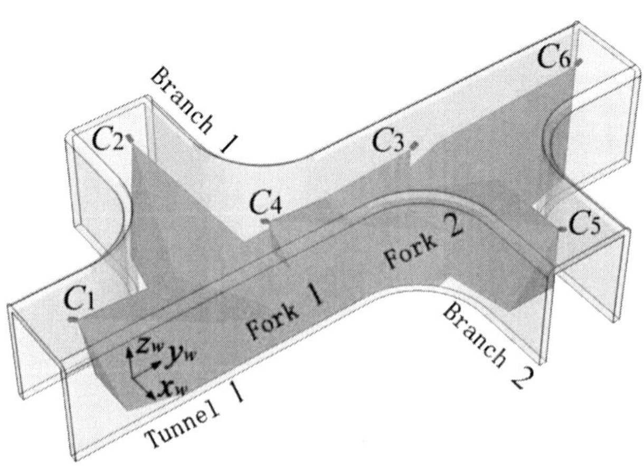


**Figure 2.** Scene of miner localization in underground coal mine with bifurcated structures.

# 3. VIDEO COLLABORATIVE LOCALIZATION OF THE MINER'S LAMP

## 3.1. The Miner's Lamp Detection with the Background Subtraction Method

In underground tunnels, generally it is difficult to distinguish miners because of coal dusts and insufficient illumination. However, miners must carry cap-lamps to offer auxiliary lighting in the underground coal mine, which provide a sharp contrast to the dim background. Thus, the background subtraction method [27] is applied in order to detect the miners' lamps reliably.

From the collected video sequences of underground tunnels, we select one frame image without miner as the initial background image. The difference image is obtained through a subtraction between the current image and the background image. In order to reduce the noise interference caused by the beam divergence of miner's lamp, the mean filter is applied to smooth the difference image obtained. An appropriate threshold is selected to turn filtered difference image into binary image which has only two pixel value, 0 and 1.

If pixel value of the binary image obtained is 0, it indicates that there is no miner in the monitoring area of a camera. Then, the background image is updated with the current image of video sequences collected by camera at the top of tunnel. In contrast, if any of the pixel values of the binary image is 1, it indicates that a miner has appeared in the monitoring area of a camera. The mapping range on the imaging plane of camera is the region with pixel value of binary image being 1, which indicates that there is a miner with cap-lamp in the area. The pixel coordinate of the mapping point of the miner's lamp on the imaging plane of camera can be obtained by calculating the geometric center of the mapping range.

## 3.2. Camera Imaging Model

The pixel coordinate of the mapping point needs to be converted to the physical coordinate in the WCS of underground coal mines. Thus, a vision imaging model of cameras is established according to the linear imaging principle of the camera [28]. Figure 3 shows the vision imaging model of cameras $C_1$, $C_2$ and $C_3$ in Figure 2.

The grey shade of Figure 3 represents the ground part of tunnels in Figure 2. *P* is the position where a miner carries a cap-lamp. $C_i$ is the camera node, assuming its position coordinate in the WCS of underground coal mine is $C_i(x_{wci},y_{wci},z_{wci})$, where the value of *i* is 1, 2 and 3. The $o_w x_w y_w z_w$ in Figure 3represents the WCS of underground coal mine that has been established in Figure 2. *OXY* represents the two-dimensional coordinate system of the imaging plane of camera. *X* axis and *Y* axis are respectively the horizontal and vertical direction on the imaging plane of camera. The pyramid composed of camera and its imaging plane represents the region that a camera covers. The Camera Coordinate System (CCS) $o_{ci}x_{ci}y_{ci}z_{ci}$of $C_i$ is established, where the position of $C_i$ is selected as the origin of its CCS, $z_{ci}$ axis is the opposite direction of optical axis of the camera, $x_{ci}$ axis and $y_{ci}$ axis are respectively same with the directions of *X* axis and *Y* axis on the imaging plane of camera.

In Figure 3, we assume that the camera $C_i$ detects a miner's lamp and calculates the pixel coordinate$P_i(u_i,v_i)$ of the mapping point on its own imaging plane. By applying the linear relationship between image resolution of the camera $C_i$ and physical size of the photosensitive element Complementary Metal Oxide Semiconductor (CMOS) in the Camera Coordinate System $o_{ci}x_{ci}y_{ci}z_{ci}$, the pixel coordinate $P_i$ $(u_i,v_i)$ of the mapping point is converted to the physical coordinate $P_{ci}(x_{ci},y_{ci},z_{ci})$ in the CCS $o_{ci}x_{ci}y_{ci}z_{ci}$ as where *f* is the focal length of camera $C_i$, *m* and *n* are the physical dimensions of the photosensitive element CMOS in the horizontal and vertical direction. The image resolution is $M \times N$, where $M$ and $N$ are respectively the number of pixels in the $X$ axis and $Y$ axis direction on the imaging plane. Expressions $m/M$ and $n/N$ are respectively the physical size of a single pixel in the $X$ axis and $Y$ axis direction on the imaging plane of camera.

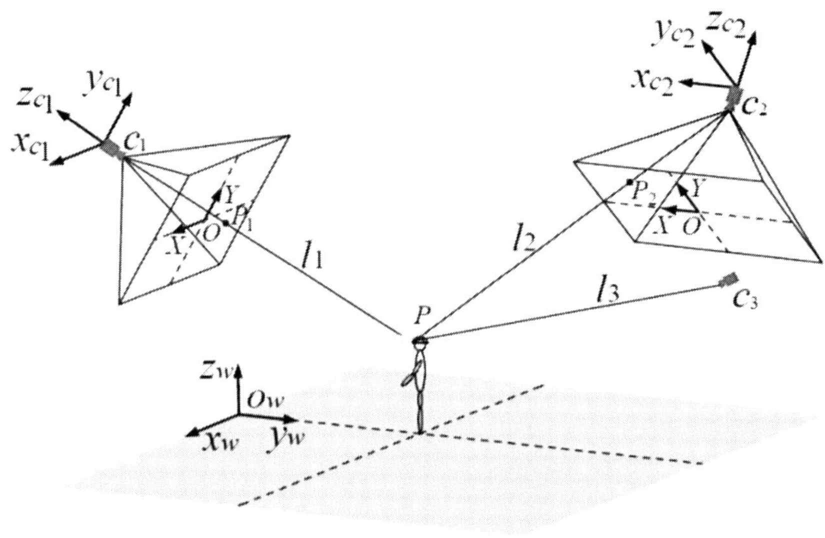


**Figure 3.** Camera vision imaging model.

$$\begin{bmatrix} x_{ci} \\ y_{ci} \\ z_{ci} \end{bmatrix} = \begin{bmatrix} m/M & 0 & -m/2 \\ 0 & -n/N & n/2 \\ 0 & 0 & -f \end{bmatrix} \begin{bmatrix} u_i \\ v_i \\ 1 \end{bmatrix} \quad (1)$$

The physical coordinate $P_{ci}(x_{ci}, y_{ci}, z_{ci})$ of the mapping point in the CCS $o_{ci}x_{ci}y_{ci}z_{ci}$ need to be converted to the physical coordinate in the WCS $o_w x_w y_w z_w$. We assume that the CCS $o_{ci}x_{ci}y_{ci}z_{ci}$ is overlapped with the WCS $o_w x_w y_w z_w$ when the WCS $o_w x_w y_w z_w$ rotate $\varphi$ degrees counterclockwise around $z_w$ axis, rotate $\theta$ degrees counterclockwise around $x_w$ axis and translate a vector $t_{wi}$ in turn. Then, the physical coordinate $P_{ci}(x_{ci},y_{ci},z_{ci})$ in the CCS $o_{ci}x_{ci}y_{ci}z_{ci}$ is turned into the physical coordinate $P_{wi}(x_{wi},y_{wi},z_{wi})$ in the WCS $o_wx_wy_wz_w$ as

$$\begin{bmatrix} x_{wi} \\ y_{wi} \\ z_{wi} \\ 1 \end{bmatrix} = \begin{bmatrix} R_{wi} & t_{wi} \\ 0_3^T & 1 \end{bmatrix} \begin{bmatrix} x_{ci} \\ y_{ci} \\ z_{ci} \\ 1 \end{bmatrix} \tag{2}$$

where the translational vector $t_{wi} = (x_{wci},y_{wci},z_{wci})$ is from the origin point to the camera $C_i$ node in the WCS of underground coal mine. Rotation matrix $R_{wi}$ is decided by the yaw angle $\varphi$ and the pitch angle $\theta$ of the camera $C_i$ as

$$R_{wi} = \begin{bmatrix} \cos\varphi & -\sin\varphi\cos\theta & \sin\varphi\sin\theta \\ \sin\varphi & \cos\varphi\cos\theta & -\cos\varphi\sin\theta \\ 0 & \sin\theta & \cos\theta \end{bmatrix} \tag{3}$$

## 3.3. Miner's Lamp Video Collaborative Localization

When a miner enters the bifurcation area of tunnels where several camera nodes have been deployed, camera $C_i$ detects the miner's lamp and calculates its pixel coordinate $P_i(u_i,v_i)$. Assume that cameras which detect the miner's lamp form a cluster by themselves in wireless way and they are in a state of time synchronization. Transforming the pixel coordinate $P_i(u_i,v_i)$ with the Equation (1), the physical coordinate$P_{ci}(x_{ci},y_{ci},z_{ci})$ of a miner's lamp mapping point in the CCS $o_{ci}x_{ci}y_{ci}z_{ci}$ can be obtained. Therefore, the physical coordinate $P_{wi}(x_{wi},y_{wi},z_{wi})$ of a miner's lamp mapping point in the WCS of underground coal mine can also be obtained by applying the Equation (2) to the physical coordinate $P_{ci}(x_{ci},y_{ci},z_{ci})$.

Establish a straight line $l_i$ between the position coordinate of camera $C_i(x_{wci},y_{wci},z_{wci})$ and the physical coordinate $P_{wi}(x_{wi},y_{wi},z_{wi})$ as

$$\frac{x_{li}-x_{wi}}{x_{wi}-x_{wci}}=\frac{y_{li}-y_{wi}}{y_{wi}-y_{wci}}=\frac{z_{li}-z_{wi}}{z_{wi}-z_{wci}}=s_i \tag{4}$$

where $(x_{li},y_{li},z_{li})$ is the coordinate of the point on the straight line $l_i$ in the WCS and $s_i$ is the parameter of the straight line $l_i$.

The three-dimension (3D) position coordinate of a miner's lamp in the WCS can be obtained by calculating the intersection of straight lines between the positions of cameras and their corresponding mapping points. However, the underground tunnels are complex and errors of camera measurement caused by position coordinate, yaw angle and pitch angle cannot be avoided. In addition, miner's lamp orientation with respect to different cameras may be different, which may lead to identification errors of miner's lamp mapping points because of the beam divergence of miner's lamp. As a result, multiple straight lines $l_i$between the positions of cameras and their corresponding mapping points generally cannot intersect at one point.

We solve this problem by finding an optimal intersection $P$ in the WCS who achieve the minimum square sum $J$ of the distance to multiple straight lines $l_i$ as

$$\begin{cases} P=\arg\min(J), J=\sum d_i^2 \\ d_i^2=(x-x_{li})^2+(y-y_{li})^2+(z-z_{li})^2 \end{cases} \tag{5}$$

The extremum problem of Equation (5) can be solved by calculating the partial derivative of $x$, $y$, $z$ and $s_i$respectively and make its value be equal to 0 as

$$\frac{\partial J}{\partial x}=0,\frac{\partial J}{\partial y}=0,\frac{\partial J}{\partial z}=0,\frac{\partial J}{\partial s_i}=0 \quad (6)$$

By this way, we can get the least square estimation of the optimal intersection $P$ which provides the position coordinate of miner's lamp in the WCS of underground coal mine. According to the 3D coordinate of a miner's lamp, the plane position coordinate $(x,y)$ of the miner in a tunnel and the altitude coordinate $z$of the miner's lamp to the ground can also be obtained.

# 4. EXPERIMENTS AND PERFORMANCES

## 4.1. Experiment in a Corridor

In order to evaluate the performances of the proposed miner's lamp video collaborative localization algorithm based on underground WMSNs, a test was carried out. A corridor with stairs shown in Figure 4was selected to simulate the bifurcation structure of underground tunnels. Lights in the corridor were turned off to simulate the insufficient illumination tunnels. Smoke machine sprayed smoke in the corridor to simulate tunnels with great amounts of water vapor and dusts. The tester wore dark blue overalls and safety helmet with miner's lamp turning on to simulate miners in underground tunnels. A World Coordinate System (WCS) was established at the entrance of the corridor with the transverse direction of the corridor as$x_w$-axis, the longitudinal direction as $y_w$-axis and the altitude direction as $z_w$-axis, respectively.

**Figure 4.** (**a**) The image of the corridor taken by camera $C_1$; (**b**) The image of the corridor taken by camera $C_2$; (**c**) The image of the corridor taken by camera $C_3$.

In the test, three isomorphic cameras TY803-130 were adopted manually at the top of the corridor by tripods. The cameras $C_1$, $C_2$ and $C_3$ were deployed according to Figure 2, and the topology of the experiment is shown in Figure 5. Cameras are wirelessly linked by Wi-Fi with carrier frequency 2.4 GHz and 802.11 n transmission protocol. Table 1 lists the intrinsic parameters of three cameras, including focal length $f$, image resolution $M \times N$ and the physical size of the photosensitive element CMOS. Table 2 lists the geometrical parameters of three cameras, including the world coordinates of the locations of cameras, the angle of yaw $\varphi$between the WCS $x_w$-axis and the Camera Coordinate System (CCS) $x_w$-axis in counterclockwise direction and the $\theta$ between the WCS $z_w$-axis and the CCS $z_w$-axis in counterclockwise direction.

**Table 1.** Intrinsic parameters of cameras.

| Parameter | Value |
|---|---|
| $f$/mm | 6.0 |
| $M \times N$ /pixel | 1280 × 960 |
| $CMOS$/mm | 4.8 × 3.6 |

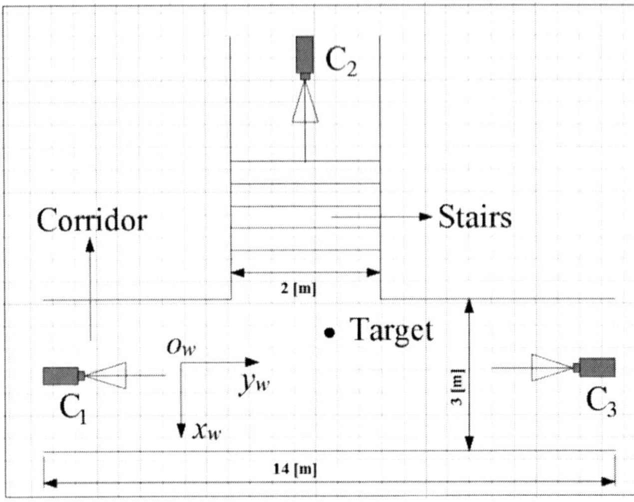


**Figure 5.** Topology of the experiment.

**Table 2.** Geometrical parameters of cameras.

| | $(x_w, y_w, z_w)$/m | $\varphi$/° | $\theta$/° |
|---|---|---|---|
| Camera 1 | (0.29, −2.20, 2.32) | 0 | 70 |
| Camera 2 | (−6.33, 2.86, 3.08) | −90 | 64 |
| Camera 3 | (0.15, 8.95, 2.34) | 180 | 70 |

The tester walked into the corridor at a normal pace. Cameras $C_1$, $C_2$ and $C_3$ worked together to monitor the corridor and collected the video information at the same time. At a certain time point, cameras $C_1$, $C_2$and $C_3$ collected the images of corridor, as is shown in Figure 4a–c, respectively.

Cameras $C_1$, $C_2$ and $C_3$ detected the miner's lamp by the difference value between the current images from their video sequences and the background images. In the experiment, 60% maximum grey value of difference image is selected as threshold. Cameras $C_1$, $C_2$ and $C_3$ detected the miner's lamp at interval of 0.5 s and calculated the pixel coordinates of mapping points of the miner's lamp on their own imaging plane. Then, the pixel coordinates were translated to the coordinates in the WCS by applying Equations (1) and (2). Therefore, three straight lines of three-dimension (3-D) between cameras $C_1$, $C_2$ and $C_3$ and their corresponding mapping points of the miner's lamp could be established in the WCS.

By applying Equations (5) and (6) to calculate the optimum intersection of the three straight lines, the position coordinate of three-dimension (3-D) of the miner's lamp in the WCS could be obtained which also provided both the plane coordinate of the tester in the corridor and the altitude of the miner's lamp from the corridor ground. 10 gauge points on the ground of the corridor had been marked and measured to show the localization precision of the miner's lamp. The tester would walk to pass these 10 gauge points and stay for one second at each gauge point. By applying the proposed algorithm, cameras $C_1$, $C_2$ and $C_3$ could locate the tester's lamp automatically at interval of 0.5 s when the tester passed each gauge point.

In order to compare the estimated localization of the tester's lamp with the actual location of the tester's lamp, the error of altitude localization is defined as

$$E_h(\mathrm{i}) = \left| h_r(\mathrm{i}) - h_l(\mathrm{i}) \right| \tag{7}$$

where $h_r(i)$ is the actual altitude of the tester's lamp from the tunnel ground and $h_l(i)$ is the altitude of the tester's lamp obtained by the proposed algorithm. Thus, the average error of the altitude localization is defined as

$$\overline{E}_h = \frac{1}{N} \sum_{i=1}^{N} E_h(i) \tag{8}$$

where $N$ is the number of gauge points in the test.

Similarly, the error of plane localization which includes errors both in $x$-axis and $y$-axis is defined as

$$E_p(i) = \sqrt{[x_r(i) - x_l(i)]^2 + [y_r(i) - y_l(i)]^2} \tag{9}$$

where (xr(i),yr(i)) is the actual plane location of the tester's lamp and (xl(i),yl(i)) is the estimated plane coordinates with the proposed algorithm. Thus, the average error of the plane localization is defined as

$$\bar{E}_p = \frac{1}{N}\sum_{i=1}^{N} E_p(i) \tag{10}$$

where $N$ is the number of gauge points in the test.

Figure 6 shows the experimental results of the altitude localization of tester's lamp with the proposed algorithm. According to the experimental data of Figure 6, the average error of the altitude localization of the tester is 3.7 cm by applying Equations (7) and (8).

Figure 7 shows the experimental results of the plane localization of tester's lamp with the proposed algorithm where the tester stood at each gauge point passed. According to the experimental data of Figure 7, the error surface of plane localization is shown in Figure 8 which applies a linear interpolation algorithm to the error of plane localization calculated with Equation (9). In Figure 8, $x$-$y$ plane is the corridor ground, $(x,y)$ coordinate represents the tester's actual location, and coordinate $z$ represents the error of plane localization between the locations estimated and the actual locations. As shown in Figure 8, the average error of the plane localization is around 10 cm with the proposed algorithm.

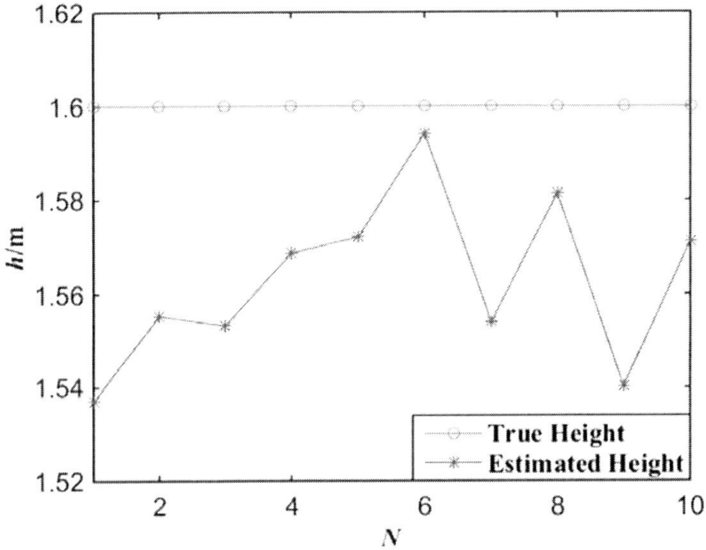


**Figure 6.** Experimental data of altitude with the miner's lamp video collaborative localization.

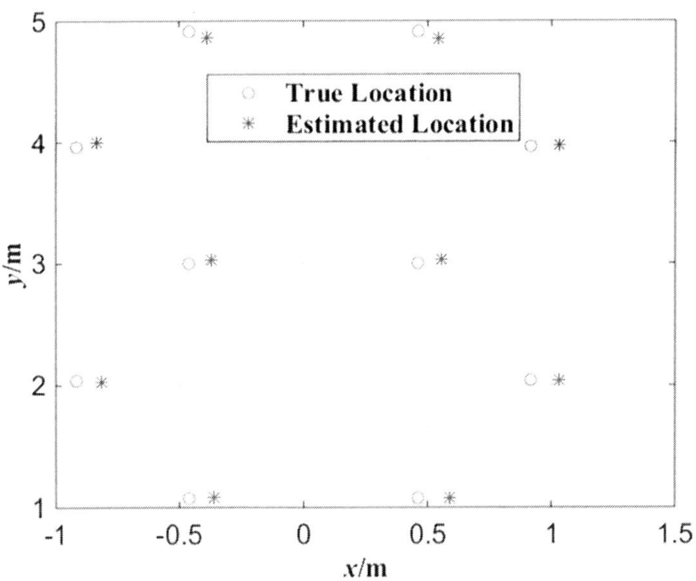


**Figure 7.** Experimental data of plane localization with the miner's lamp video collaborative localization.

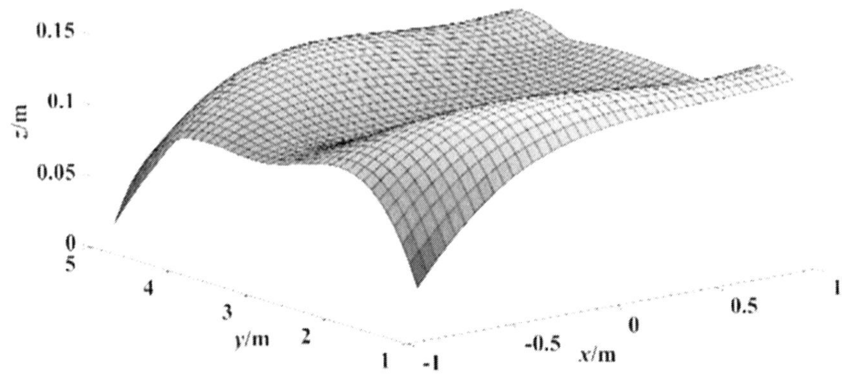


**Figure 8.** Error surface of plane localization with the miner's lamp video collaborative localization.

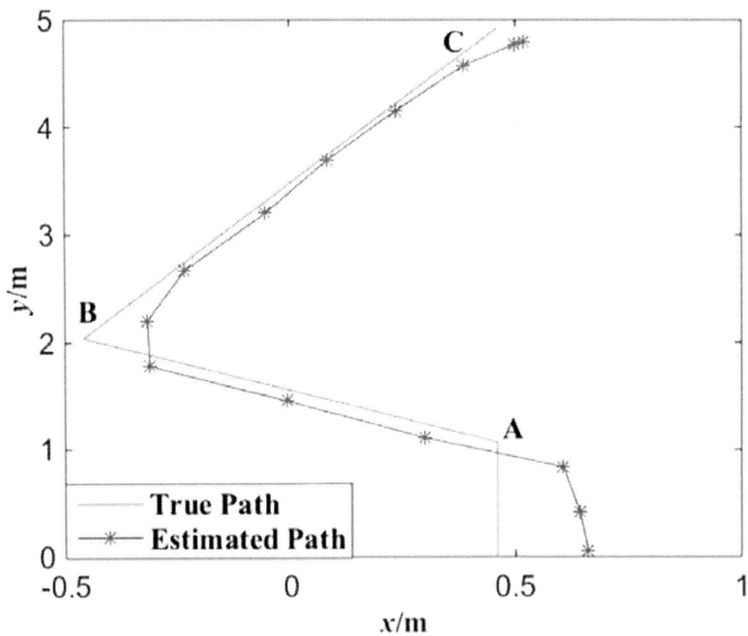


**Figure 9.** Trajectory tracked with the miner's lamp video collaborative localization.

In order to evaluate the performance of the miner's lamp collaborative localization algorithm under the condition that the miner moves continuously, cameras $C_1$, $C_2$ and $C_3$ were also applied to detect the movement of the tester in the corridor. The tester walked straight into the corridor from the entrance at a normal pace. Then, the tester turned about 45° at point A and walked about 1.5 m to point B. After that, the tester turned again at point B and walked toward point C. Figure 9 shows the tester's actual motion trajectory as well as the estimated motion trajectory with the proposed algorithm at interval of 0.5 s. InFigure 9, the motion trajectory estimated is obtained by connecting two adjacent tester's locations with line segments.

In the bottom right of Figure 9, it can be observed that the location error is larger when the tester began to enter into the corridor. This is because when the tester began to enter into the corridor, the mapping points of tester's cap-lamp were respectively on the edge of the imaging planes of cameras $C_1$, $C_2$ and $C_3$ at the same time. Due to the non-linear relationship between object and image on the imaging edge [29], the mapping points of tester's cap-lamp on the imaging planes will distort to some extent which will lead to larger location error. However, the maximum error for the movement orbit tracing is about 20 cm which is also acceptable to trace a miner. In contrast, the location error is smaller when the tester was around central area of the corridor. This is because the images of cameras $C_1$, $C_2$ and $C_3$ have a good linear relationship when the miner's lamp is around central area which results in smaller errors.

## 4.2. Experiment in an Underground Tunnel

In order to evaluate the performances of the proposed algorithm in the realistic scenario of underground tunnels, a test was also carried out in the realistic scenario of an underground tunnel. The test was performed in the underground tunnel of Track Vibration Abatement and Control Laboratory at Beijing Jiaotong University, as shown in Figure 10. The width and the altitude of the underground tunnel are 4 m with horseshoe shaped structure. In the test, six isomorphic cameras TY803-130 were deployed manually at the top of the tunnel by tripods. The cameras $C_1$, $C_2$, $C_3$, $C_4$, $C_5$ and $C_6$ were powered by lithium batteries.

Figure 10 and Figure 11 show the actual deployment of cameras and the topology of the experiment, respectively. A World Coordinate System (WCS) was established at the entrance of the tunnel with the transverse direction of the tunnel as $x_w$-axis, the longitudinal direction as $y_w$-axis and the altitude direction as $z_w$-axis, respectively. As shown in Figure 11, the six isomorphic cameras were divided into three groups and the two cameras in the same group were deployed oppositely, with cameras $C_1$ and $C_3$ as the first group, cameras $C_2$ and $C_5$ as the second group and cameras $C_4$ and $C_6$ as the third group. Cameras $C_1$ and$C_3$ are deployed crosswise along the direction of center line of the bended tunnel to monitor the area of tunnel entrance together. In the same way, cameras $C_2$ and $C_5$ are deployed crosswise along the direction of center line of the bended tunnel to monitor the area of bended tunnel together. Similarly, cameras $C_4$ and $C_6$are deployed crosswise along the longitudinal direction of the straight tunnel to monitor the area of straight tunnel together.

The intrinsic parameters of six cameras, including focal length $f$, image resolution $M \times N$ and the physical size of the photosensitive element CMOS are listed in Table 1. Table 3 lists the geometrical parameters of six cameras, including the world coordinates of the locations of cameras, the angle of yaw $\varphi$ between the WCS$x_w$-axis and the CCS $x_w$-axis in counterclockwise direction and the $\theta$ between the WCS $z_w$-axis and the CCS$z_w$-axis in counterclockwise direction.

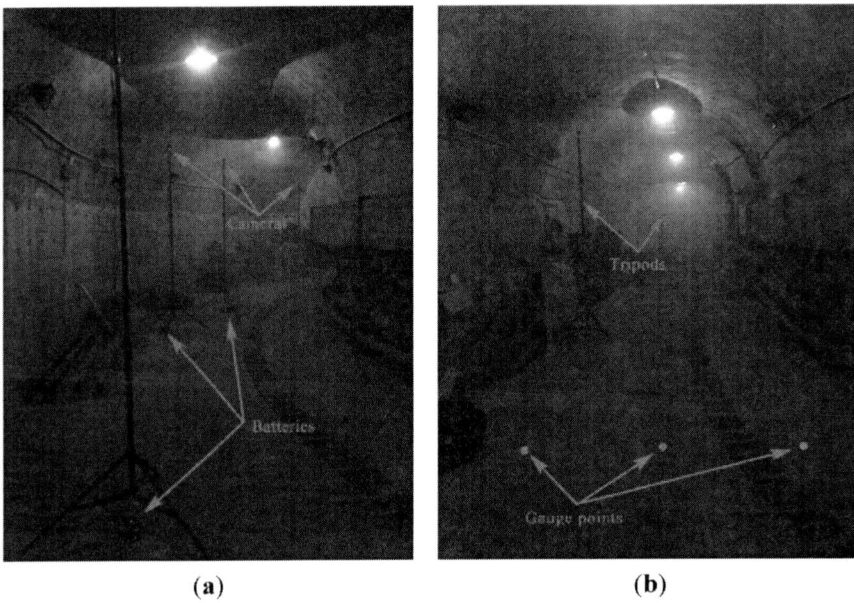


**Figure 10.** (**a**) The actual deployment of cameras; (**b**) Gauge points on the ground of the tunnel.

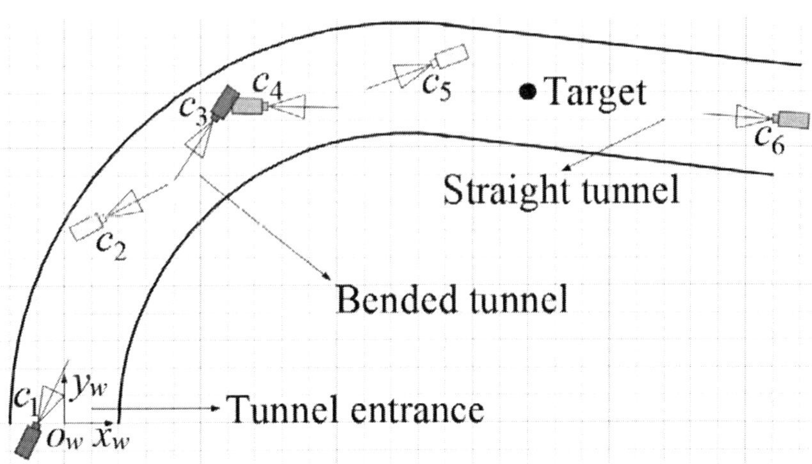


**Figure 11.** Topology of the experiment.

**Table 3.** Geometrical parameters of cameras.

| | $(x_w, y_w, z_w)$/m | $\varphi$/° | $\theta$/° |
|---|---|---|---|
| Camera 1 | (−1.00, 0, 2.87) | −26.6 | 68.0 |
| Camera 2 | (1.50,7.50, 2.90) | −62.7 | 67.4 |
| Camera 3 | (5.50, 11.00, 2.88) | 146.1 | 70.6 |
| Camera 4 | (7.50, 11.50, 2.90) | −91.2 | 71.8 |
| Camera 5 | (13.50, 13.00, 2.87) | 110.1 | 71.2 |
| Camera 6 | (26.00, 11.00, 2.85) | 86.0 | 72.1 |

In the test, tester wore dark blue overalls and safety helmet with miner's lamp turning on to simulate miners in underground tunnels. The tester walked into the tunnel at a normal pace. Cameras $C_1$, $C_2$, $C_3$, $C_4$,$C_5$ and $C_6$ worked together to monitor the tunnel and collected the video information at the same time.Figure 12a–f showed the images of the tunnel collected by cameras $C_1$, $C_2$, $C_3$, $C_4$, $C_5$ and $C_6$ at a certain time point, respectively.

Cameras $C_1$ and $C_3$, cameras $C_2$ and $C_5$, and cameras $C_4$ and $C_6$ detected the miner's lamp by the difference value between the current images from their video sequences and the background images, respectively. In the experiment, 60% maximum grey value of difference image is selected as threshold. Cameras $C_1$, $C_2$, $C_3$, $C_4$, $C_5$ and $C_6$ detected the miner's lamp at interval of 0.5 s and calculated the pixel coordinates of mapping points of the miner's lamp on their own imaging plane. Then, the pixel coordinates were translated to the coordinates in the WCS by applying Equations (1) and (2).

Therefore, when two cameras of a certain group detected the miner's lamp simultaneously, two straight lines of three-dimension (3-D) between two cameras in the group and their corresponding mapping points of the miner's lamp could be established in the WCS. By applying Equations (5) and (6) to calculate the optimum intersection of the two straight lines, the position coordinate of three-dimension (3-D) of the miner's lamp in the WCS could be obtained which provided both the plane coordinate of the tester along the tunnel and the altitude of the miner's lamp from the tunnel ground.

In order to evaluate the localization precision of the proposed algorithm when applied to the realistic scenario of underground tunnel, dozens of gauge points had been marked on the ground of the tunnel, as shown in Figure 10b. Gauge points formed grids with horizontal and vertical spacing of one meter (1 m) on the ground of the tunnel.

In the test, the tester walked and passed these gauge points. The tester would stand and squat about one second (1 s) at each gauge point passed, respectively. The coordinates of gauge points in the WCS and the altitude of the miner's lamp from the tunnel ground when the tester stood and squatted had been measured manually which provided actual locations of tester's lamp. When the tester could be located by cameras of two groups simultaneously, the average value of localizations of the two groups was selected as the localization result of the tester's lamp.

Figure 14 shows the experimental results of the plane localization of tester's lamp with the proposed algorithm where the tester stood at each gauge point passed. Note that the intersection of grids in Figure 14is the gauge point with 1 m spacing. We can note that most of the location points estimated have errors in the same direction compared with the gauge points in Figure 14. This is mainly because that the tester's lamp was always toward the opposite direction of tunnel entrance while walking in the test. According to the experimental data of Figure 14, the error of plane localization for up-right posture of the tester can be obtained by applying Equation (9) as shown in Figure 15. When the number of gauge points is 83, the average error of the plane localization for up-right posture of the tester is 27.4 cm by applying Equation (10). Similarly, when the error of plane localization is less than 50 cm, there will be no obvious deviation for the estimation of plane localization or trajectory of miners along tunnels. The experimental results of Figure 14and Figure 15 indicate that the plane coordinates of the tester on the ground of a tunnel can be precisely estimated with the proposed miner's lamp video collaborative localization algorithm. It is worth mentioning that it can be ensured to recognize a miner's lamp by adjusting the threshold effectively. However, since the illumination of miner's lamp is generally intensive in different background illumination such as in Figure 4and Figure 10, 60% maximum grey value of difference image was selected as the threshold, which provided a good performance of recognition and localization accuracy in both scenarios.

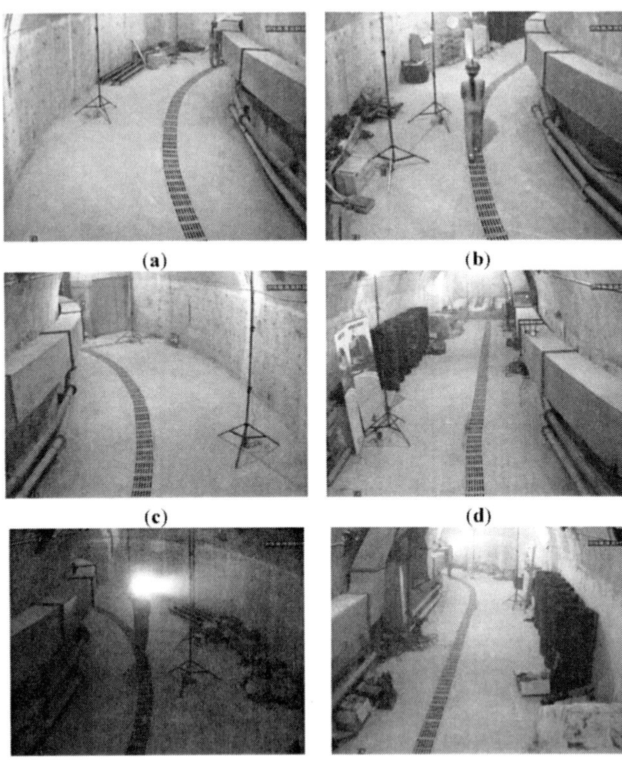


**Figure 12.** (**a**) The image of the tunnel taken by camera $C_1$; (**b**) The image of the tunnel taken by camera $C_2$; (**c**) The image of the tunnel taken by camera $C_3$; (**d**) The image of the tunnel taken by camera $C_4$; (**e**) The image of the tunnel taken by camera $C_5$; (**f**) The image of the tunnel taken by camera $C_6$.

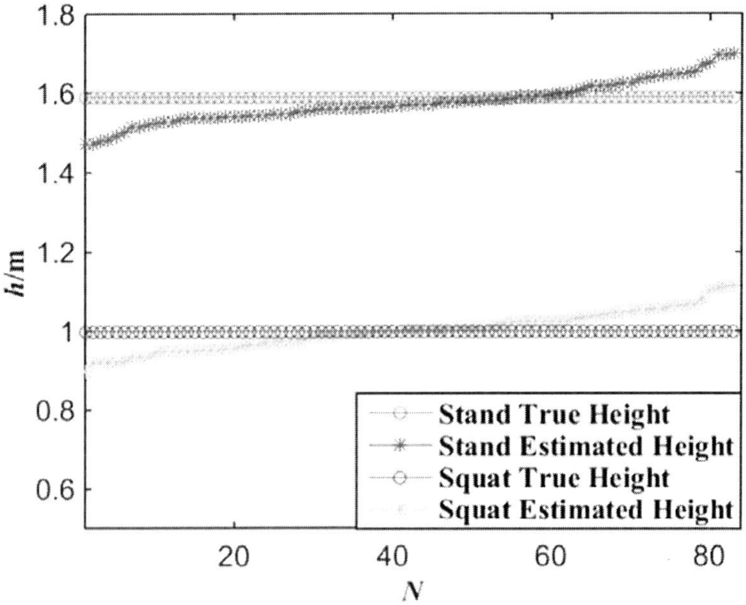


**Figure 13.** Experimental data of altitude with the miner's lamp video collaborative localization.

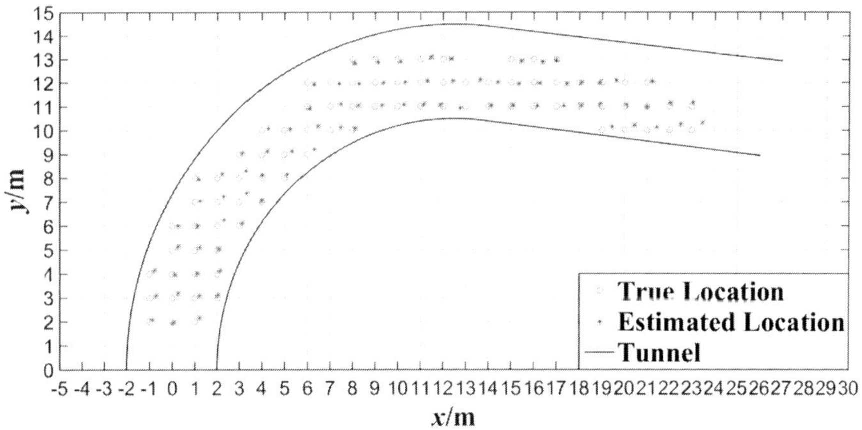


**Figure 14.** Experimental data of plane localization with the miner's lamp video collaborative localization.

In order to evaluate the performance of the proposed algorithm under the condition that miner moves continuously, cameras $C_1$, $C_2$, $C_3$, $C_4$, $C_5$ and $C_6$ were also applied to detect the movement of the tester in the tunnel. The tester walked along a 90° broken line and a curved line in the tunnel at a normal pace, respectively. Figure 16 shows the tester's actual motion trajectory along the 90° broken line as well as the estimated motion trajectory with the proposed algorithm at interval of 0.5 s. Figure 16 shows the tester's actual motion trajectory along the curved tunnel as well as the estimated motion trajectory with the proposed algorithm at interval of 0.5 s. In Figure 16 and Figure 17, the motion trajectory estimated is obtained by connecting two adjacent tester's locations with line segments. From Figure 16 and Figure 17, it can be observed that the motion trajectory of a miner can be estimated accurately by locating the miner continuously with the proposed miner's lamp video collaborative localization algorithm. By this way, a miner can be tracked along tunnels.

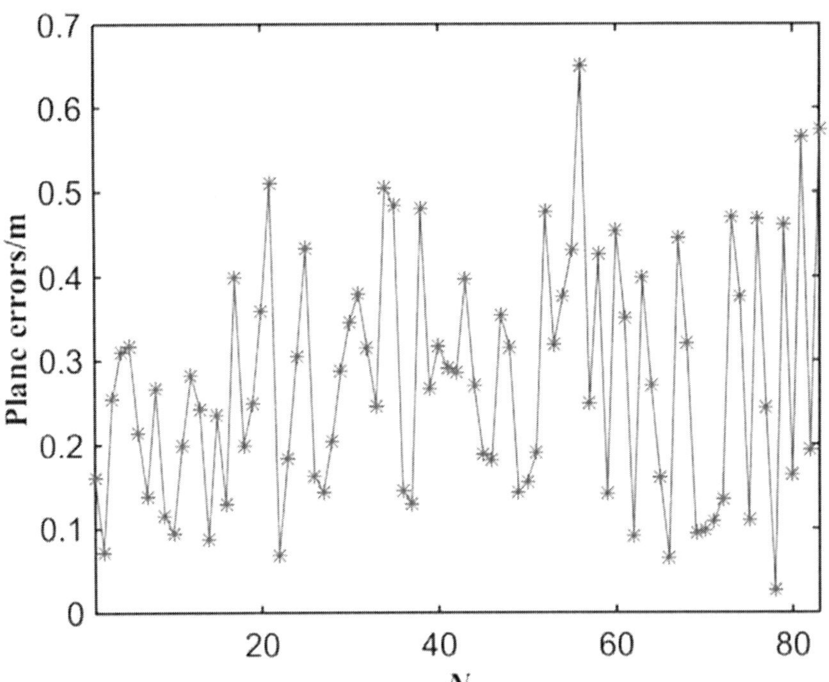


**Figure 15.** The error of plane localization with the miner's lamp video collaborative localization.

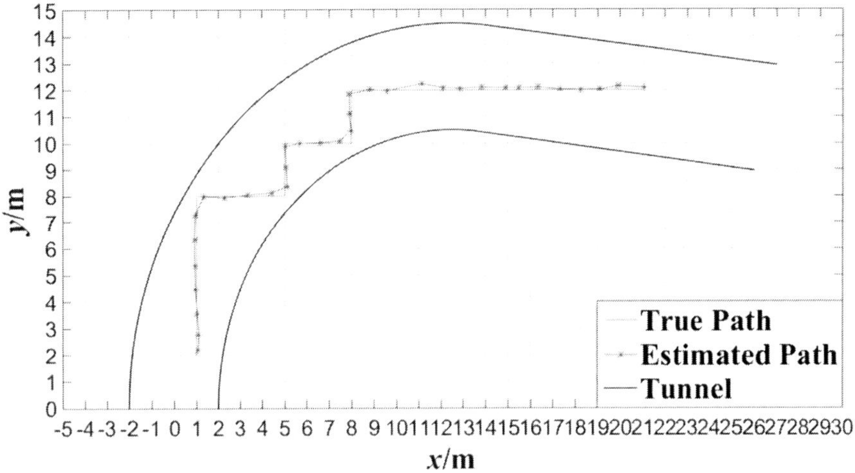


**Figure 16.** Trajectory of broken line tracked with the miner's lamp video collaborative localization.

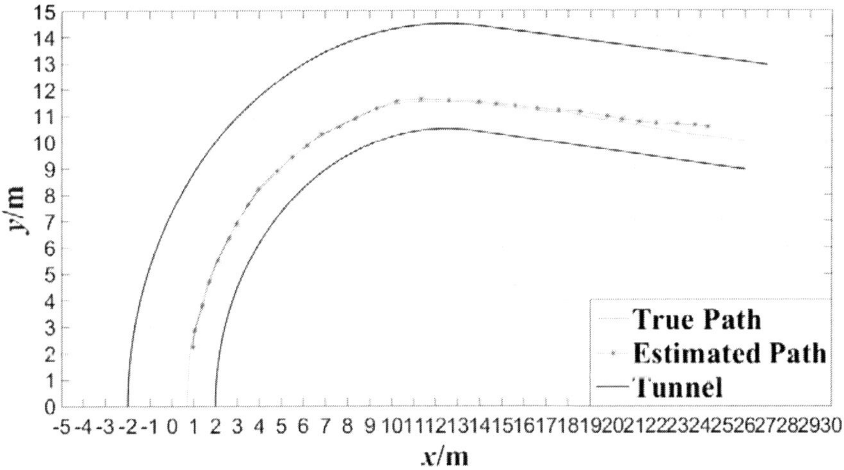


**Figure 17.** Trajectory of curved line tracked with the miner's lamp video collaborative localization.

# 5. CONCLUSIONS

Wireless multimedia sensor networks (WMSNs) will improve the monitoring performance of underground coal mines significantly.

Based on WMSNs deployed in underground coal mine, a miner's lamp video collaborative localization algorithm was proposed to locate miners in the scene of bifurcated structures of underground tunnels.

1. To detect a miner's lamp, it is proposed to apply background difference method to get the difference value between the background image of underground tunnels and the current image of the video sequences collected.

2. A least squares method is proposed to find the optimal intersection which solves the problem that multiple straight lines between the positions of cameras and their corresponding mapping points generally cannot intersect at one point.

3. The experimental results in a corridor indicate that the average error of the altitude localization is 3.7 cm, and the average error of the plane localization is about 10 cm with the proposed algorithm. The experimental results in an underground tunnel indicate that the average error of the altitude localization for up-right and squat postures of the tester are 4.7 cm and 3.9 cm respectively, and the average error of the plane localization for up-right posture of the tester is 27.4 cm with the proposed algorithm.

## ACKNOWLEDGMENTS

This work was supported by the Natural Science Foundation of China under Grant (51474015) and the National Science & Technology Pillar Program (2013BAK06B03).

## AUTHOR CONTRIBUTIONS

The work was carried out in collaboration between all authors. Kaiming You, Wei Yang and Ruisong Han conceived and designed the experiments. Wei Yang contributed materials. Kaiming You performed the experiments. Kaiming You and Wei Yang analyzed the data. Kaiming You wrote the paper. Each author contributed with expertise in their domain. All authors have contributed to, seen and approved the manuscript.

## REFERENCES

1. Xu, J.; Yang, W.; Zhang, L.; Han, R.; Shao, X. Multi-Sensor Detection with Particle Swarm Optimization for Time-Frequency Coded Cooperative WSNs Based on MC-CDMA for Underground Coal Mines. *Sensors* **2015**, *15*, 21134–21152.

2. Zhang, Y.; Yang, W.; Han, D.; Kim, Y.I. An Integrated Environment Monitoring System for Underground Coal Mines—Wireless Sensor Network Subsystem with Multi-Parameter Monitoring.*Sensors* **2014**, *14*, 13149–13170.

3. Pei, Z.; Deng, Z.; Xu, S.; Xu, X. Anchor-Free Localization Method for Mobile Targets in Coal Mine Wireless Sensor Networks. *Sensors* **2009**, *9*, 2836–2850.

4. Sun, J. Technologies of Monitoring and Communication in the Coal Mine. *J. China Coal Soc.* **2010**, *35*, 1925–1929. (In Chinese).

5. Yang, W.; Feng, X.; Cheng, S.; Sun, J. The Theories and Key Technologies for the New Generation Mine Wireless Information System. *J. China Coal Soc.* **2004**, *29*, 506–509. (In Chinese).

6. Almalkawi, I.T.; Zapata, M.G.; Al-Karaki, J.N.; Morollo-Pozo, J. Wireless Multimedia Sensor Networks: Current Trends and Future Directions. *Sensors* **2010**, *10*, 6662–6717.

7. Akyildiz, I.F.; Melodia, T.; Chowdhury, K.R. A Survey on Wireless Multimedia Sensor Networks.*Comput. Netw.* **2007**, *51*, 921–960.

8. Akyildiz, I.F.; Melodia, T.; Chowdhury, K.R. Wireless Multimedia Sensor Networks: Applications and Testbeds. *Proc. IEEE* **2008,** *96,* 1588–1605.

9. Zhao, X.; Wang, Z.; Zhao, K. Research on Distributed Image Compression Algorithm in Coal Mine WMSN. International Journal of Digital Content Technology and Its Applications. *Adv. Inst. Converg. Inf. Technol.* **2011**, *5*, 283–291.

10. De, D.; Gupta, M.D.; Sen, A. Energy Efficient Target Tracking Mechanism Using Rotational Camera Sensor in WMSN. *Proced. Technol.* **2012**, *6*, 674–681.

11. Qi, M.; Zhang, R.; Jiang, J.; Li, X. Moving Object Localization with Single Camera Based on Height Model in Video Surveillance. In Proceedings of the 1st International Conference on Bioinformatics and Biomedical Engineering, Wuhan, China, 6–8 July 2007; pp. 490–493.

12. Öztarak, H.; Akkaya, K.; Yazici, A. Lightweight Object Localization with a Single Camera in Wireless Multimedia Sensor Networks. In Proceedings of the Global Telecommunications Conference, Honolulu, HI, USA, 30 November–4 December 2009; pp. 1–6.

13. Zhang, B.; Luo, H.; Liu, J.; Zhao, F.; Liu, S.; Lin, Y. Passive Target Localization Algorithm for Multimedia Sensor Networks. *J. Southeast Univ. Nat. Sci. Edit.* **2011,** *41,* 266–269. (In Chinese).

14. Liu, L.; Zhang, X.; Ma, H. Optimal Node Selection for Target Localization in Wireless Camera Sensor Networks. *IEEE Trans. Veh. Technol.* **2010,** *59,* 3562–3576.

15. Liu, L.; Ma, H.; Zhang, X. Collaborative Target Localization in Camera Sensor Networks. In Proceedings of the IEEE Wireless Communications and Networking Conference, Las Vegas, NV, USA, 31 March–3 April 2008; pp. 2403–2407.

16. Li, W.; Portilla, J.; Moreno, F.; Liang, G.; Riesgo, T. Improving Target Localization Accuracy of Wireless Visual Sensor Networks. In Proceedings of the 37th IEEE Industrial Electronics Conference, Melbourne, Australia, 7–10 November 2011; pp. 3814–3819.

17. Lin, Q.; Zeng, X.; Jiang, X.; Jin, X. Video Nodes Based Collaborative Object Localization Framework in Wireless Multimedia Sensor Networks. *Equip. Manuf. Technol. Autom.* **2011**, *317*, 1078–1083.

18. Karakara, M.; Qi, H. Collaborative Localization in Visual Sensor Networks. *ACM Trans. Sens. Netw.***2014**, *10*, 127–146.

19. Boulanouar, T.; Lohier, S.; Rachedi, A.; Roussel, G. A Collaborative Tracking Algorithm for Communicating Target in Wireless Multimedia Sensor Networks. In Proceeding of the Wireless and Mobile Networking Conference, Vilamoura, Portugal, 20–22 May 2014; pp. 1–7.

20. Chilgunde, A.; Kumar, P.; Ranganath, S.; Huang, W. Multi-Camera Target Tracking in Blind Regions of Cameras with Non-Overlapping Fields of View. In Proceeding of the British Machine Vision Conference (BMVC), London, UK, 7–9 September 2004; pp. 397–406.

21. Sun, J.; Jia, N. Human Target Matching and Visual Tracking Method in Coal Mine. *J. China Univ. Min. Technol.* **2015**, *44*, 525–533. (In Chinese).

22. Pescaru, D.; Curiac, D.I. Anchor Node Localization for Wireless Sensor Networks Using Video and Compass Information Fusion. *Sensors* **2014**, *14*, 4211–4224.

23. Tron, R.; Terzis, A.; Vidal, R. *Distributed Consensus Algorithms for Image-Based Localization in Camera Sensor Networks*; Springer London: London, UK, 2011; pp. 289–302.

24. Zhang, B.; Liu, J.; Luo, H. Monocular Vision-Based Distributed Node Localization for Wireless Multimedia Sensor Networks. In Proceedings of the 2nd International Congress on Image & Signal, Tianjin, China, 17–19 October 2009; pp. 1–5.

25. Elhamifar, E.; Vidal, R. Distributed Calibration of Camera Sensor Networks. In Proceedings of the Third ACM/IEEE International Conference on Distributed Smart Cameras, Como, Italy, 30 August–2 September 2009; pp. 1–7.

26. Barton-Sweeney, A.; Lymberopoulos, D.; Savvides, A. Sensor Localization and Camera Calibration in Distributed Camera Sensor Networks. Broadband Communications. In Proceedings of the 3rd International Conference on Networks and Systems, San Jose, CA, USA, 1–5 October 2006; pp. 1–10.

27. Piccardi, M. Background Subtraction Techniques: A Review. *IEEE Int. Conf. Syst. Man Cybern.* **2004**, *4*, 3099–3104.

28. Heikkilä, J.; Silvén, O. A Four-Step Camera Calibration Procedure with Implicit Image Correction. In Proceedings of IEEE Computer Society Conference on Computer Vision and Pattern Recognition, Puerto Rico, Territory, 17–19 January 1997; pp. 1106–1112.

29. Douxchamps, D.; Chihara, K. High-Accuracy and Robust Localization of Large Control Markers for Geometric Camera Calibration. *IEEE Trans. Pattern Anal. Mach. Intel.* **2009**, *31*, 376–383.

# CHAPTER 4

# An Underground Air-Route Temperature Prediction Model for Ultra-Deep Coal Mines

**Shuai Zhu [1], Shiyue Wu [1,*], Jianwei Cheng [2,*], Siyuan Li [2] and Mingming Li [3]**

[1] *College of Mining Engineering, Taiyuan University of Technology, Taiyuan 030024, China*
[2] *College of Safety Engineering, China University of Mining and Technology, Xuzhou 221116, China*
[3] *Department of Mining Engineering, West Virginia University, Morgantown, WV 26505, USA*

## ABSTRACT

Due to modern mining methods deployed in recent years, production of coal mines has been expanded significantly compared to thirty years ago. As a consequence, the mining depth of coal mines is becoming ever deeper. A common world-wide problem that underground coal mines are currently experiencing is the hazard caused by the underground hot environment, which also promotes a great need of reliable mitigation measures to assist mine operators controlling the heat stress for miners as well as maintaining the normal operation of the mine. In this paper, a model for underground air-route temperature prediction in ultra-deep mines based on previous findings was developed. In developing this model, the idea of heat balance was used to establish the temperature calculation equation. Various underground heat sources (air compress, wall oxidation, underground heat, machinery, *etc.*) are covered in the model to improve the prediction accuracy. In addition, a PC-based numerical tool was also developed to aid users using such a mathematical model. Finally, a few temperature measurements for an ultra-deep underground coal mine were performed to demonstrate the applicability of the proposed mathematical prediction model.

## KEYWORDS

underground air temperature prediction; mining environment control; computer program software

## 1. INTRODUCTION & BACKGROUND INFORMATION

Mining works over a long period have been going ever deeper for the extraction of mineral resources. One of the serious problems associated with depth is the heat produced as a result of increasing geothermal gradient, air compression, and ground water. According to the Chinese government official report released in 2007 [1], among the state owned coal mines, 28.5% of coal mines in China have a depth over 600 m and from more than 70 mines temperatures in working faces exceed 26 °C and 30 of them are over 30 °C. In addition, the highest temperature in an underground working face can achieve 37 °C [1]. Currently, the depth of a typical coal mine is increasing at a rate of more than 20 m annually in China [2]. The hot environment in deep underground coal mines in China causes serious problems for mine operators. The same thing occurs in other countries as well. For example, the air temperature in western South Africa underground gold mines rises to 50 °C at a depth of 3300 m. The temperature in the Fengyu lead-zinc ore mine in Japan is up to 80 °C at a depth of 500 m [3].

The thermal stress due to the high temperature in deep mines affects not only the health, but also the safety performance of miners. Employees working in deep mines are exposed to a high temperature and coupled with strenuous work causes dehydration. Such working conditions can lead to various levels of heat disorders, including heat cramps, heat exhaustion, heat collapse, and heat stroke. The most serious and dangerous one is heat stroke which can be fatal or cause multiple organ failures. Heat can also affect a worker's attention, hand-to-eye coordination, mental acuity, and other neurological functions, which contribute to mine accidents [4]. The work efficiency of persons working in a hot, humid working environment is reduced as work must be carried out at a slower pace to avoid the body overheating and additionally, dexterity and coordination, a person's ability to observe irregular faint optical signs, to stay alert and to make quick decisions are adversely affected by the thermal strain. Studies of a 3-h drilling operation conducted under varying room temperatures show that the best results were achieved at 84 °F (29 °C), and the performance was reduced to 75% capacity at a room temperature of 91 °F (33 °C), to 50% at 96 °F (35 °C), and to 25% at 99 °F (37 °C). The study also shows that the upper limit for the best performance of a novice is 77 °F (25 °C) [5].

A prediction of the underground air temperature is required to make heat control plans effectively and economically. In order to control heat in underground mines, it is important to calculate or estimate the amplitude of the heat flow from each source. The total heat gain will be matched by the air-cooling requirements. From administrative perspectives, the prediction of the underground air temperature will help to make a selection and acclimatization plan for workers working in hot and humid work areas. The South Africa mining industry has made significant progress in reducing the incidence of heat strokes through prior selection and acclimatization of miners. Acclimatization is a long-term adjustment of an individual to stress. An acclimatized person can perform many tasks in a hot and humid work site whereas a non-acclimatized person cannot work. For example, a person should be given enough time to adapt to a hot work site where the wet bulb globe temperature (WBGT) may exceed 79 °F (26 °C). A recommended six-day acclimatization schedule calls for miners to work in a hot work site for 50% of the time on the first work day, and an additional 10% of time on the days that follow the first day [5].

The control of heat in mines can be realized through a combination of ventilation, refrigeration, and cooling. Bossard [6] provided ways to calculate the magnitude of the heat flow gained from heat sources to the mine air in an underground mine. Torres and Singh studied thermal stress and human feelings in underground mining [7]. Cong *et al.*, designed an air cooling system using a low-temperature water spay to control heat hazard in mines [8]. He [3] developed a High Temperature Exchange Machinery System (HEMS) for deep mine heat hazard control under the principle of extracting cold energy from mine water inrush. This cooling system was applied in several deep coal mines in China. It firstly extracts cold energy from mine water inrush to cool working faces; then uses the heat extracted by HEMS to supply heat to buildings and bath water instead of using a boiler. Wang, *et al.* [9] proposed research to evaluate the thermal status of the human body by considering parameters from both the human body and the surrounding environment. Wagner studied the major methods of cooling deep and ultra-deep mines [10]. His research showed that the deciding factor behind the choice of surface or underground cooling of ventilation air is the auto-compression of ventilation air. The application of chilled service water together with ice slurry was found to be the most cost effective way of mine cooling by Wagner. The amount of water required for mine cooling was reduced by using ice slurry as the cooling medium. Therefore, the cost of pumping water to the underground working faces was reduced.

In summary, the proper prediction of the temperature in underground working environments is significant for mine operation and plays important roles such as:

- Reasonable assessment of the heat stress for underground miners
- Proper use of mitigation measures against thermal damages to improve working efficiency;
- Providing accurate fundamental data for calculating the cooling load once a system is designed;
- Efficient selection of the cooling medium and reduction of the refrigeration cost;

Therefore, to have better heat control in underground mines, accurate air temperature is the key for mine operators. Due to the complex environmental conditions in underground mines, the air temperature varies in deep mines due to several types of heat transfer processes. In this paper, a model for predicting the underground air-route temperature for ultra-deep mines based on previous findings is developed. In addition, a numerical tool is also developed to facilitate the calculations. The following sections start by introducing the basic mathematical models which make up the backbone of the computer program. Then, the flowchart is described and a user-interface is developed for implementing the algorithm. Finally, a comparison with the measured data shows that this model has a wide applicability and could be applied in operations.

# 2. THE MATHEMATICAL MODEL

## 2.1. Overview of Underground Heat Sources

Major heat sources in underground mines which are capable of modifying the environment conditions include: Auto-compression, wall rock, underground water, machinery, and light. Other potential sources of heat in mines include human metabolism, oxidation, blasting, rock movement, and pipelines [11].

### 2.1.1. Auto-Compression

When air flows from the surface down a mine shaft, it is compressed and heated in the same way as occurs in a compressor. Auto-compression occurs when the potential energy is converted to thermal energy [11]. The temperature increase due to auto-compression can be up to 10 °C per kilometer of vertical depth, whether in a shaft or in a decline [12]. The heat from air auto-compression would be more significant in deeper mines as it increase with the vertical depth. Also it is obvious that mines sunk in hot areas are hot work sites due to the effect of hot surface air combining with the auto-compression.

### 2.1.2. Wall Rock

The underground mines at deeper depth encounter hotter levels of the earth crust due to the geothermal gradient. The geothermal gradient of the upper crust is generally between 15 and 40 °C/km [12]. The temperature difference between rock and air cause the heat flow from rock to air, which is the major source of the heat source in underground mines. A survey has shown that in deep mines, heat transfer from the rock mass contributes more than 75% of total mine heat load [10]. The heat transfer rate is primarily dependent on the thermal properties of rock and the temperature difference between the rock and air. Another key factor is moisture as moisture lowers heat transfer resistance at the interface and lowers air dry-bulb temperature [11].

### 2.1.3. Underground Water

There are two different water sources encountered in underground mines: ground water and mine water. The water temperature approaches or even exceeds that of the rock as water and heat are derived from the surrounding rock or geothermal sources [11]. The heat transfer from water to mine air is mainly through evaporation, which increases the latent heat of the air.

### 2.1.4. Machinery and Lighting

Nearly all the energy consumption of machinery underground adds heat to the mine air. The power losses and most of the work done are converted directly to heat or indirectly through friction [11]. The commonly used power equipment in mines is electrical, compressed-air, or internal-combustion (diesel) machinery. The heat produced by diesel powered equipment is equivalent to about 90% of the calorific value of the total fuel consumed. The electrical substations, fans, and non-submersible pumps all dissipate a portion of the electrical energy supplied to them as heat to the mine air. Almost all of the power of the conveyor belts is consumed in overcoming friction between the belts and moving material on the horizontal. The heat of friction is transferred to the mine atmosphere [6].

## 2.2. Assumptions Made in the Model

The following assumptions were made in developing the underground air-route temperature prediction program model:

1. The shape of all underground roadways is assumed to be circular;
2. The underground rock is isotropic homogenous material;
3. Compared to the radial temperature gradient, the axial temperature gradient can be negligible in rock;
4. A uniform air temperature distribution is assumed on the same cross section of roadway;
5. The virgin-rock temperature is assumed be uniform everywhere at the same elevation;

6. There is no air-leakage occurring underground, which means travelling air volume quantity is conserved anywhere within a specified tunnel.

## 2.3. Descriptions of Mathematical Model

When surface air is introduced into a mine, both the temperature and humidity are affected. The contributing factors include:

1. Variations of temperature and humidity on the surface;
2. Heat and water exchange between the surrounding rock of mine and ventilation air;
3. Air auto-compression heat;
4. Oxidation heat from ore body, coal, and timbers;
5. Heat from electrical and mechanical equipment and other sources.

Within these influence factors, the first one can be observed with the aid of historical data. The third, fourth and fifth heat sources are nothing to do with the air temperature. Hence, they are easily calculated. The second heat source is related to the temperature and the humidity of the ventilation air. It may bring some difficulties when deriving results.

According to the law of heat conservation, the heat change of ventilation air in an underground tunnel should be equal to the summation of external heat changed over a period of time. The setup of the mathematical equation is:

$$\sum Q = G(i_2 - i_1) \tag{1}$$

where $\Sigma Q$ is the summation of heat generated within a specified underground tunnel, J/h; $G$ is the ventilation air mass, kg/h; $i_1$ and $i_2$ are the enthalpy of air when inflowing and outflowing from the specified underground tunnel.

Generally speaking, once the initial condition is known, $i_2$ can be calculated by Equation (1). In addition, the air temperature can be determined by using thermodynamics equations. Hence, it can be seen that the fundamental of air temperature prediction is to establish a quasi-steady heat transfer and balance the equations.

## 2.4. Heat from the Mine Shaft

Very similar to the way in which gas in a compressor responds, air entering a mine through a shaft is compressed and heated as it flows downward. Air auto-compression occurs when potential energy is converted to thermal energy [11]. However, on considering the interchange of heat and water content in the shaft, the temperature may not increase at 1 °C/100 m as for theoretical calculations [13]. Therefore, based on observation statistics in some Chinese mines, the following equation was used to calculate the heat due to the compression:

$$Q_{Shaft} = -\ 0.976G\,(273 + t_1)\left[\left(1 + \frac{0.0124h}{101.325 + 0.012H}\right)^{0.286} - 1\right] \qquad (2)$$

where: $Q_{Shaft}$ is the heat transfer in the mine shaft, J/h; $G$ is the ventilation air mass, kg/h; $t_1$ is the air temperature at the surface, °C; $h$ is the height of the mine shaft, $m$. $H$ is the elevation of the mine shaft head,$m$.

## 2.5. Heat from Wall Rock

The temperature of subsurface rock rises steadily with depth. The so-called virgin-rock temperature is the most important factor affecting the subsurface temperature. It keeps at a constant value between about

15–50 m beneath the surface. Then, it increases with depth at approximately a uniform rate (or geothermal gradient). The virgin-rock temperature can be calculated as:

$$t_{Rock-1} = t_{Rock-0} + \frac{H_{Rock-0} - H_{Rock-1}}{G_R} \quad (3)$$

where: $t_{Rock-1}$ is the virgin-rock temperature at subsurface elevation point 1, °C; $t_{Rock-0}$ is the constant temperature beneath the surface, °C; $H_{Rock-0}$ is the elevation for constant temperature, *m*; $H_{Rock-1}$ is the elevation at subsurface elevation point 1, *m*; $G_R$ is the geothermal gradient, *m*/°C.

After opening roadways underground, due to the temperature difference between the virgin-rock and airflow, heat transfer from the wall rock to the ventilation air occurs. However, this process is not only unsteady but also complex. The amount of heat transferred is highly dependent on ventilation time, rock properties, *etc.* By applying thermodynamic analyses, the following equation is used to calculate the heat transfer [13]:

$$Q_{Rock} = k_{\tau} U L (\bar{t}_{Rock} - \bar{t}_{Air}) \quad (4)$$

where: $Q_{Rock}$ is the heat transfer from the wall rock to ventilation air in unit time, J/h; $k_{\tau}$ is the unsteady heat transfer coefficient, J/(m$^2$·h·°C); The value of $k_{\tau}$ is highly dependent on the ventilation time and can be determined by Equation (5); *U* is the perimeter of the underground roadway, *m*; *L* is the length of underground roadway, *m*; t¯Rock is the average virgin-rock temperature at the beginning and end of the roadway, °C; t¯Rock is the average air temperature at the beginning and end of the roadway, °C.

To justify the unsteady heat transfer coefficient $k_\tau$, depending on the ventilation time, it can be determined as:

When the ventialtion time is less than 10 years

$$K_\tau = \begin{cases} \frac{1}{1+\frac{\lambda g}{2\alpha R}} \cdot \left[ \frac{\lambda_g}{2R} + \frac{b}{2\sqrt{\tau}\left(1+\frac{\lambda_g}{2\alpha R}\right)} \right] \\ 0.5 \frac{\lambda g^{0.65} (C_g \Upsilon_g)^{0.2} \alpha^{0.15}}{R^{0.4} \tau^{0.2}} \end{cases} \quad (5)$$

When the ventialtion time is more than 10 years

where: $k\tau$ is the unsteady heat transfer coefficient, J/(m²·h·°C); $\lambda g$ is the coefficient of rock thermal conductivity, J/(m·h·°C); α is the heat transfer coefficient of wall rock, J/(m²·h·°C); it can be calculated as $\alpha \sim = \sim \frac{2\varepsilon G^{0.8} U^{0.2}}{S}$, where ε is the roughness of rock wall; $G$ is the ventilation air mass, kg/h; $U$ is the perimeter of underground roadway, $m$; $S$ is the cross-sectional area of underground roadway, $m^2$; $R$ is the hydraulic radius of underground roadway, $m$; $R = 0.564\sqrt{S}$; $b$ is rock heat storage coefficient, J/(m²· h^0.5· °C); $b = 2\sqrt{\frac{\lambda_g C_g \Upsilon_g}{\varepsilon}}$, where $C_g$ is the rock specific heat, J/(kg·°C); $\gamma_g$ is the rock density, kg/m³; τ is the ventilation time, hours.

## 2.6. Heat from Roadway Wall Oxidation

Oxidation processes involving the mineral, backfill, and timber oxidized with oxygen in mines could contribute heat to the mine air [11]. It can be a very important source of heat [14]. The heat amount for the oxidation, which is considered for the wall oxidation process, can be calculated by:

$$Q_{Oxidation} = ULq_{Wall} \tag{6}$$

where: $Q_{Oxidation}$ is the oxidation heat transfer from the wall, J/h; $q_{Wall}$ is the coefficient wall oxidation of heat transfer, J/(m$^2$·h); Table 1 lists the values for different types of roadways; $U$ is the perimeter of the underground roadway, *m*; $L$ is the length of the underground roadway, *m*.

**Table 1.** Recommended values for different $q_{Wall}$.

| Roadway Type | $q_{Wall}$ J/(m$^2$·h) |
|---|---|
| Mine working face | $6.28 \times 10^4$ |
| Intake or return of working face | $1.46 \times 10^4$ |
| Rock drift or rock roadway support by roofbolts or shotcrete | $1.57 \times 10^4$ |
| Mine mains or coal drift | $2.72 \times 10^4$ |

## 2.7. Heat from Underground Machinery

Underground machinery, such as the working face shear, belt conveyor, *etc.* can convert electrical power into heat. Then the air temperature could be increased. Calculation of the heat gain from machinery proceeds from a knowledge of the total power and load factors, which are listed as:

$$Q_{Machinery} = \frac{\phi_1 \phi_2 \phi_3 \tau_R A}{24} N \tag{7}$$

where: $Q_{Machinery}$ is the heat generated from machinery, J/h; $\phi_1$ is the installation factor (ratio of motor maximum consumption power to rated power), usually taken around 0.7; $\phi_2$ is the simultaneous factor (ratio of summation of used motor rated power to the total rated power); $\phi_3$ is the load factor, usually taken around 0.4–0.5; $\tau_R$ is the machinery running time per day, h/d; $A$ is the conversion factor, $A = 3.6 \times 10^6$ J/h; $N$is the machinery rated power, kW.

The equation to calculate the heat generated by an auxiliary fan in an underground mine development head is:

$$Q_{A.Fan} = \frac{80GP_F}{\eta_M \eta_F} \tag{8}$$

where: $Q_{A.Fan}$ is the heat generated from the auxiliary fan, J/h; $G_{Air}$ is the ventilation air mass, kg/h; $P_F$ is the auxiliary fan working pressure, Pa; $\eta_M$ is the motor efficiency; $\eta_F$ is the fan efficiency

## 2.8. Heat from Wall Rock in Underground Working Face

The heat from wall rock in an underground working face is similar to the heat from wall rock in a normal roadway. Hence, a similar equation is used to calculate the heat:

$$Q_{W.Face} = \alpha_F U_W L_W (\bar{t}_{Rock} - \bar{t}_{Air}) \tag{9}$$

where: $Q_{W.Face}$ is the heat transfer from the wall rock around the working face to ventilation air in unit time, J/h; $\alpha_F$ is the heat transfer coefficient of wall rock around the working face, J/(m₂·h·°C); it can be calculated as $\alpha_F = 5.3 + 3.6v$, where $v$ is the air velocity at the working face, m/s; $U_w$ is the perimeter of the working face, *m*; $L_w$ is the length of the working face, *m*; t¯Rock is the average virgin-rock temperature at the beginning and end of the working face, °C; t¯Air is the average air temperature at the beginning and end of the working face, °C.

## 2.9. Temperature Calculation Using Thermodynamic Equations

The enthalpy of air, which is the sum of the enthalpies of dry air and water vapor, can be expressed as:

$$i = 1.005t + 0.001d(25016 + 1.884t) \qquad (10)$$

where: $i$ is the enthalpy of moist air, Kg/m$^3$; $t$ is the temperature, °C; $d$ is the specific humidity, kg/kg;

The equation of specific humidity (weight of water vapor contained per unit weight of dry air) can be calculated as:

$$d=0.622\frac{\phi P_{q,b}}{B - \phi P_{q,b}} \qquad (11)$$

where: $d$ is the specific humidity of moist air, kg/m$^3$; $B$ is the barometric pressure, Pa; $\varphi$ is the relatively humidity, %; $P_{q,b}$ is the saturated vapor pressure, Pa.

The saturation vapor pressure $P_{q,b}$ is the pressure of a vapor when it is in equilibrium with the liquid phase. It is solely dependent on the temperature. The empirical equation to calculate $P_{q,b}$ is shown as (when the temperature is between 0–200 °C):

$$P_{q,b} = \exp(\frac{C_1}{T} + C_2 + C_3T + C_4T^2 + C_5T^3 + C_6ln(T)) \qquad (12)$$

where: $P_{q,b}$ is the saturated vapor pressure, Pa; T is the absolute temperature, K; $C_{1\text{-}5}$ are coefficients, which are equal to −5800.2206;

0.41764768 × $10^{-4}$; 1.3914993; −0.14452093 × $10^{-7}$; −0.04860239 and 6.5459673, respectively.

The barometric pressure at an elevated underground level can approximately be estimated as:

$$B = \frac{101325}{\left(1 + \frac{H}{44300}\right)^{5.256}} \tag{13}$$

where: $B$ is the barometric pressure, Pa; $H$ is the elevation below the seal level, $m$.

The relatively humidity relationship between two points is:

$$\phi_2 = \phi_1 + k_\phi L_{1-2} \tag{14}$$

where: $\varphi_1$ and $\varphi_2$ are the relatively humilities at point 1 and 2, %; $L_{1-2}$ is the distance between two points, $m$;$k_\varphi$ is the relatively humidity gradient, 1m. The value of $k_\varphi$ is highly dependent on the underground environmental condition. In most cases, it can be determined by field measurements or historical statistics.

Density of moist air can be calculated by Equation (15):

$$\rho{=}0.00348\frac{B}{T} - 0.001315\frac{\phi \cdot P_{q,b}}{T} \tag{15}$$

where: ρ is the density of moist air, kg/m$^3$; T is the absolute temperature, K; $B$ is the barometric pressure, Pa; φ is the relatively humidity; $P_{q,b}$ is the saturated vapor pressure, Pa.

## 3. CALCULATIONS STRATEGIES FOR DIFFERENT UNDERGROUND ROADWAYS

Underground roadways are connected with each other to form a network. Different roadways serve different functions in coal mining production. In consideration of heat source distribution in the subsurface and of the characteristics that underground roadways possess, the following four conditions can be identified:

- The mine shaft: it is a very unique tunnel for underground coal mining because the spatial form of it is vertical. Generally, comparing to the great number of underground roadways that a mine has, there may only be a few shafts serving for hoist and ventilation purposes. The biggest heat sources for the shaft are the heat from wall rock and air auto-compression. However, due to watering, the heat from wall rock is used up by a water evaporation process. Hence, air auto-compression is the only significant heat source.

- Underground horizontal or inclined roadways: There are several heat sources that can affect the air temperature increasing, which include heat from wall rock, underground water, oxidation, machinery, air compression, *etc.* In most cases, only the heat from wall rock and wall oxidation are important factors [15]. Therefore, these two are considered in this program model.

- Underground development head: this roadway is similar to the above one. However, the major difference is its dead-end opening. Artificial ventilation must be applied in order to circulate the air for diluting contaminants. Hence, an auxiliary fan is used and the heat generated by fan must be considered.

- Coal working face: The coal is cut by a longwall shearer in this roadway. Once the coal contacts oxygen, the oxidation process can bring an amount of heat. In addition, the shearer, conveyor or other electrical equipment can also be heat sources to increase the air temperature.

# 4. ALGORITHM IMPLEMENTATION

## 4.1. States of the Numerical Tool

In order to avoid time-consuming and tedious calculation work, a feasible computer program needs to be developed. With the help of using Visual Basic computer language, the authors computerized the mathematical model into an underground air-route temperature prediction software program. Users can easily input the data into the featured visualized program interface. Then, the calculation can be automatically performed and the finial prediction results can be outputted in tabular format and exported into a Microsoft Excel spreadsheet. Figure 1 shows the screenshot of the main interface of the underground temperature prediction program.

When the program is run, the main interface is shown as below:

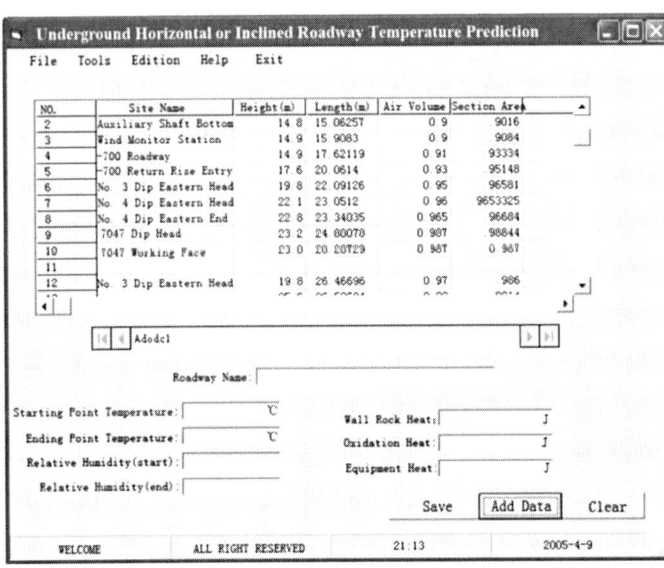


**Figure 1.** Main interface of underground temperature prediction program.

## 4.2. The Flow Chart

As mentioned in "Calculations Strategies" in the previous section, different types of underground roadways may consider different underground heat sources. Hence, different calculation procedures should be implemented. In the developed underground air-route temperature prediction program, four types of underground roadways have been organized, which are the mine shaft, the underground horizontal or inclined roadway, the underground development head and the coal working face. A corresponding flow chart of the program is listed as in Figure 2. It can be seen that the trial-and-error method is used in the program. The temperature prediction result could be outputted once convergence is met.

# 5. APPLICATION CASE STUDY

## 5.1. Information about a Typical Coal Mine

To examine the reliability of the developed mathematical model and developed software program, a typical ultra-deep underground mine was chosen as the study problem. This mine is located in the central part of China. The geology survey shows that the constant temperature of the subsurface rock is 16 °C with a depth of 30 m. The average geothermal gradient of the coal mine is 3.24 °C/100 m. Currently, the mining production elevation level is −700 m. Field measurement of the rock virgin temperature is already as high as 37.70 °C. In addition, the next production level is planned at −980 m. It can be seen that such a high temperature environment could bring serious problems to miners' health and safety as well as production efficiency.

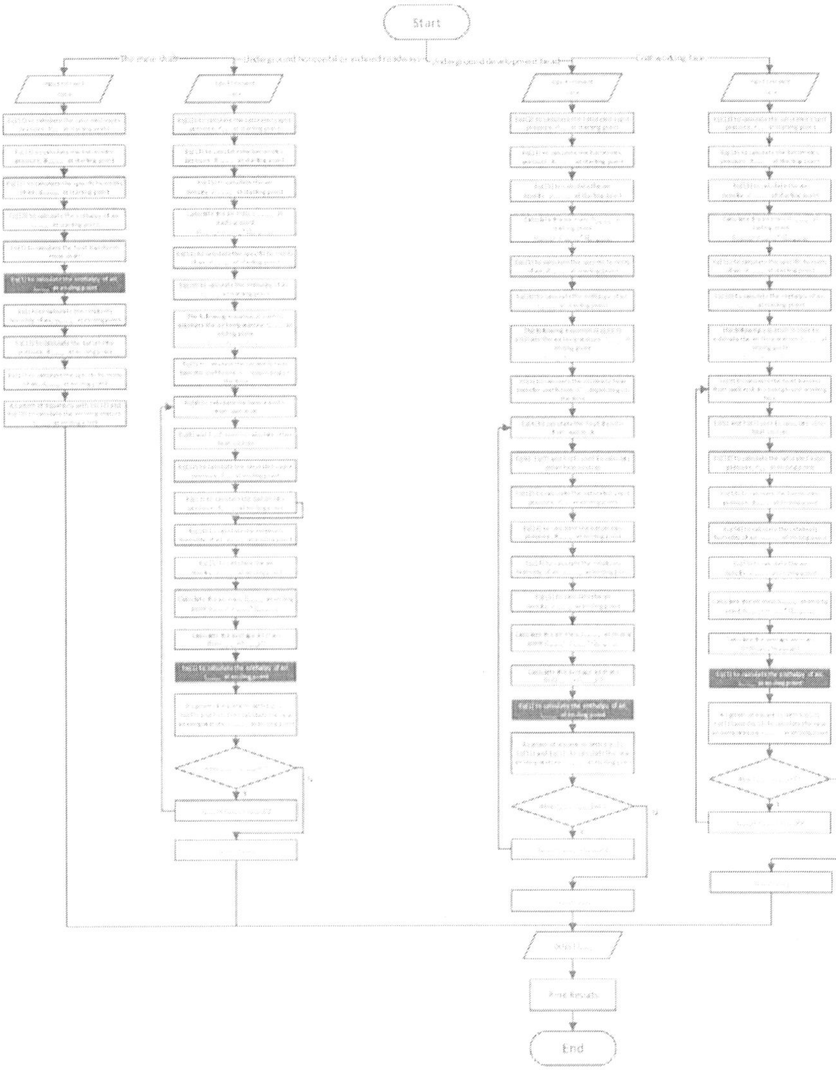


**Figure 2.** Calculation flow chart of the program.

## 5.2. Air-Route Descriptions and Graphical Representation

In this study, a total of three air-routes were chosen to validate the reliability of the prediction results. InFigure 3, the distribution of prediction points is presented. The routes are designed by following the air flow direction until the end including the underground working face and development head, which is described as:

- **Rout <1>:** 1 (Surface) → 2 (Mine shaft bottom) → 3 (Airflow observation station) →4 (−700 North level airflow split point) →11 ($-700_{-4}$ level roadway) → 12 ($-700_{-4}$ level roadway) → 13 (7407 air-intake entry) → 14 (7407 working face head ) → 15 (7407 working face end)
- **Rout <2>:** 1 (Surface) → 2 (Mine shaft bottom) → 3 (Air measuring station) →4 (−700 level airflow split point) → 21 ($-700_{-3}$ level roadway) → 22 ($-700_{-3}$ level roadway) → 23 ($-700_{-3}$ level roadway) → 24 (7301 air-intake entry) → 25 (7301 working face head ) → 26 (7301 working face end)
- **Rout <3>:** 1(Surface) →2 (Mine shaft bottom) → 3 (Air measuring station) → 31 (−700 South level airflow split point) → 32 (−700 level roadway) → 33 (−700 level roadway)→ 34 (Dip head entry in south area) → 35 (Dip head entry in south area) → 36 (Mine development head)

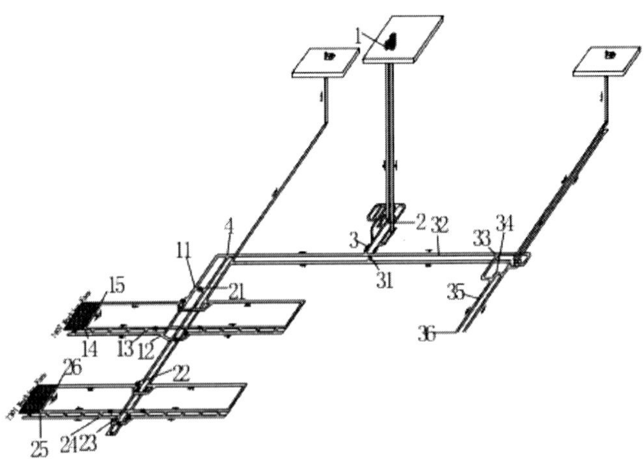


**Figure 3.** The prediction route and distribution of measurement points.

## 5.3. Prediction Results and Comparisons

Table 2, Table 3 and Table 4 list the input parameters when using the computer program to calculate the predicted air temperature at each prediction point for the three routes. Table 5, Table 6 and Table 7 show the derived results from the computer program. In addition, in order to check the calculation reliability, the in-mine field temperature measurements are also included for comparison reasons. Figure 4 shows the calculated difference between predicted and field measured temperature. It can be seen that the biggest difference is 1.1 °C. Most results agree well with one other. Hence, this mathematical model has better accuracy and reliability.

# 6. CONCLUSIONS

Accurate evaluation and prediction of the air temperature increase in underground coal mines can provide very valuable information for mine operators in order to take preventive measures for mitigating thermal problems for miners as well as improving their health. In order to deal with such a problem, by consideration of various heat sources and their roles in increasing the temperature, a mathematical model was proposed for providing air-temperature prediction in mines and a numerical tool and interface was also developed. This program provides a versatile, reliable, and easy-to-use tool for users who care about or operate in underground ventilation management.

**Table 2.** Input parameters for route <1>.

| No. | Parameter Names | Symbol in Equations | Unit | 1 | 2 | 3 | 4 | 11 | 12 | 13 | 14 | 15 |
|---|---|---|---|---|---|---|---|---|---|---|---|---|
| | | | | Site Name | | | | | | | | |
| | | | | Mine Shaft | Mine Shaft Bottom | Airflow observation Station | −700 North Level Airflow Split Point | $-700_{-4}$ Level Roadway | $-700_{-4}$ Level Roadway | 7407 Air-Intake Entry | 7407 Working Face Head | 7407 Working Face End |
| | | | | Parameter Values | | | | | | | | |
| 1 | Ventilation air mass | $G$ | kg/h | 8185 | - | - | - | - | - | - | - | - |
| 2 | Coefficient of rock thermal conductivity | $\lambda_g$ | J/(m·h·°C) | 9627 | 7534 | 9836 | 9836 | 9836 | 9836 | 9836 | 987 | 987 |
| 3 | Toughness of rock wall | $\varepsilon$ | mm | 1 | 1.5 | 1.5 | 1.5 | 1.5 | 1.5 | 2.0 | 2.5 | 3.0 |
| 4 | The perimeter of underground roadway | $U$ | m | 32.1 | 12.5 | 10.1 | 10.1 | 9.6 | 9.6 | 8.7 | 13.2 | 13.2 |
| 5 | The cross-sectional area of underground roadway | $S$ | $m^2$ | 19.5 | 15.3 | 12.6 | 12.6 | 11.8 | 11.8 | 10.9 | 15.2 | 15.2 |
| 6 | The rock specific heat | $C_g$ | J/(kg·°C) | 837 | 826 | 852 | 852 | 852 | 852 | 852 | 1046 | 1046 |
| 7 | Rock density | $\gamma_g$ | kg/m³ | 1709 | 2652 | 2567 | 2567 | 2567 | 2567 | 2567 | 1568 | 1568 |
| 8 | Ventilation time | $\tau$ | year | 15.6 | 14.8 | 13.8 | 5.6 | 4.8 | 2.8 | 1.8 | 1 | 1 |
| 9 | The length of underground roadway | $L$ | m | 780 | 10 | 120 | 450 | 160 | 251 | 60 | 360 | 167 |
| 10 | Coefficient wall oxidation of heat transfer | $q_{Wall}$ | J/(m²·h) | $2.8 \times 10^4$ | $1.3 \times 10^4$ | $1.3 \times 10^4$ | $1.7 \times 10^4$ | $1.7 \times 10^4$ | $1.9 \times 10^4$ | $2.3 \times 10^4$ | 5.7 | $7.2 \times 10^4$ |
| 11 | machinery running time per day | $\tau_R$ | h/d | - | - | - | 16 | - | - | 16 | 16 | 16 |
| 12 | Machinery rated power | $N$ | kw | | | | | | | 125 | 126 | 1300 |
| 13 | Air quantity | $Q$ | m³/min | 6789 | 4956 | 4985 | 2584 | 2236 | 2156 | 1716 | 1600 | 1600 |
| 14 | Air velocity in working face | $v$ | m/s | - | - | - | - | - | - | - | - | 3.5 |
| 15 | Absolute elevation | $H$ | m | −710 | −710 | −705 | −708 | −715 | −715 | −709 | −712 | −713 |

**Table 3.** Input parameters for route <2>.

| No. | Parameter names in Equations | Symbol | Unit | 1 | 2 | 3 | 4 | 21 | 22 | 23 | 24 | 25 | 26 |
|---|---|---|---|---|---|---|---|---|---|---|---|---|---|
| | | | | Site Name | | | | | | | | | |
| | | | | Mine Shaft | Mine Shaft Bottom | Airflow Observation Station | −700 North Level Airflow Split Point | $-700_{-3}$ Level Roadway | $-700_{-3}$ Level Roadway | $-700_{-3}$ Level Roadway | 7301 Air-Intake Entry | 7301 Working Face Head | 7301 Working Face End |
| | | | | Parameter Values | | | | | | | | | |
| 1 | Ventilation air mass | $G$ | kg/h | 8185 | - | - | - | - | - | - | - | - | - |
| 2 | Coefficient of rock thermal conductivity | $\lambda_g$ | J/(m· h· °C) | 9627 | 7534 | 9836 | 9836 | 9836 | 9836 | 9836 | 9836 | 987 | 987 |
| 3 | Toughness of rock wall | $\varepsilon$ | mm | 1 | 1.5 | 1.5 | 1.5 | 1.5 | 1.5 | 2.0 | 2.0 | 2.5 | 3.0 |
| 4 | The perimeter of underground roadway | $U$ | m | 32.1 | 12.5 | 10.1 | 10.1 | 9.6 | 9.6 | 8.7 | 8.7 | 13.2 | 13.2 |
| 5 | The cross-sectional area of underground roadway | $S$ | $m^2$ | 19.5 | 15.3 | 12.6 | 12.6 | 11.8 | 11.8 | 10.9 | 10.9 | 15.2 | 15.2 |
| 6 | The rock specific heat | $C_g$ | J/(kg· °C) | 837 | 826 | 852 | 852 | 852 | 852 | 852 | 852 | 1046 | 1046 |
| 7 | Rock density | $\gamma_g$ | kg/m$^3$ | 1709 | 2652 | 2567 | 2567 | 2567 | 2567 | 2567 | 2567 | 1568 | 1568 |
| 8 | Ventilation time | $\tau$ | year | 15.6 | 14.8 | 13.8 | 5.6 | 5.1 | 3.8 | 1.9 | 1.9 | 1 | 1 |
| 9 | The length of underground roadway | $L$ | m | 780 | 10 | 120 | 370 | 460 | 132 | 60 | 60 | 286 | 163 |
| 10 | Coefficient wall oxidation of heat transfer | $q_{Wall}$ | J/(m$^2$· h) | $2.8 \times 10^4$ | $1.3 \times 10^4$ | $1.3 \times 10^4$ | $1.7 \times 10^4$ | $1.7 \times 10^4$ | $1.9 \times 10^4$ | 2.3*10$^4$ | $2.3 \times 10^4$ | $5.7 \times 10^4$ | $7.2 \times 10^4$ |
| 11 | machinery running time per day | $\tau_R$ | h/d | - | - | - | 16 | - | - | 16 | 16 | 16 | 16 |
| 12 | Machinery rated power | $N$ | kw | | | | | | | 125 | 125 | 126 | 1300 |
| 13 | Air quantity | $Q$ | m$^3$/min | 5789 | 4956 | 4985 | 2401 | 2159 | 2013 | 1517 | 1517 | 1400 | 1400 |
| 14 | Air velocity in working face | $v$ | m/s | - | - | - | - | - | - | - | - | - | 3.5 |
| 15 | Absolute elevation | $H$ | m | −710 | −710 | −705 | −708 | −718 | −718 | −720 | −720 | −723 | −723 |

**Table 4.** Input parameters for route <3>.

| No. | Parameter Names | Symbol | Unit | 1 | 2 | 3 | 31 | 32 | 33 | 34 | 35 | 36 |
|---|---|---|---|---|---|---|---|---|---|---|---|---|
| | | | | Site Name | | | | | | | | |
| | | | | Mine Shaft | Mine Shaft Bottom | Air Measuring Station | −700 South Level Airflow Split Point | −700 Level Roadway | −700 Level Roadway | Dip Head Entry in South Area | Dip Head Entry in South Area | Mine Development Head |
| | | | | Parameter Value | | | | | | | | |
| 1 | Ventilation air mass | $G$ | kg/h | 8185 | - | - | - | - | - | - | - | - |
| 2 | Coefficient of rock thermal conductivity | $\lambda_g$ | J/(m·h·°C) | 9627 | 7534 | 9836 | 9836 | 9926 | 9926 | 9926 | 997 | 997 |
| 3 | Toughness of rock wall | $\varepsilon$ | mm | 1 | 1.5 | 1.5 | 1.5 | 1.5 | 2.0 | 3.0 | 3.5 | 5.0 |
| 4 | The perimeter of underground roadway | $U$ | m | 32.1 | 12.5 | 10.1 | 9.6 | 9.6 | 8.7 | 8.7 | 6.2 | 6.2 |
| 5 | The cross-sectional area of underground roadway | $S$ | $m^2$ | 19.5 | 15.3 | 12.6 | 11.8 | 11.8 | 10.9 | 10.9 | 9.4 | 9.4 |
| 6 | The rock specific heat | $C_g$ | J/(kg·°C) | 837 | 826 | 852 | 852 | 852 | 852 | 852 | 1046 | 1046 |
| 7 | Rock density | $\gamma_g$ | kg/m$^3$ | 1709 | 2652 | 2567 | 2567 | 2567 | 2567 | 2567 | 1568 | 1568 |
| 8 | Ventilation time | $\tau$ | year | 15.6 | 14.8 | 13.8 | 3.1 | 1.8 | 1.1 | 0.8 | 0.5 | 0.2 |
| 9 | The length of underground roadway | $L$ | m | 780 | 10 | 120 | 25 | 260 | 120 | 260 | 60 | 10 |
| 10 | Coefficient wall oxidation of heat transfer | $q_{Wall}$ | J/(m$^2$·h) | $2.8 \times 10^4$ | $1.3 \times 10^4$ | $1.3 \times 10^4$ | $1.7{*}10^4$ | $1.9{*}10^4$ | $2.3 \times 10^4$ | $2.3 \times 10^4$ | $5.7 \times 10^4$ | $8.2 \times 10^4$ |
| 11 | machinery running time per day | $\tau_R$ | h/d | - | - | - | - | - | - | - | 16 | 16 |
| 12 | Machinery rated power | $N$ | kw | - | - | - | - | - | - | - | 126 | 120 |
| 13 | Air quantity | $Q$ | m$^3$/min | 6789 | 4956 | 1986 | 1986 | 1986 | 1986 | 880 | 460 | 460 |
| 14 | Absolute elevation | $H$ | m | −710 | −710 | −705 | −788 | −788 | −812 | −812 | −825 | −825 |
| 15 | The auxiliary fan working pressure | $P_F$ | Pa | | | | | | | | 600 | 3200 |
| 16 | The motor efficiency | $\eta_M$ | % | - | - | - | - | - | - | - | 70 | 82 |
| 17 | The fan efficiency | $H_F$ | % | | | | | | | | 65 | 82 |

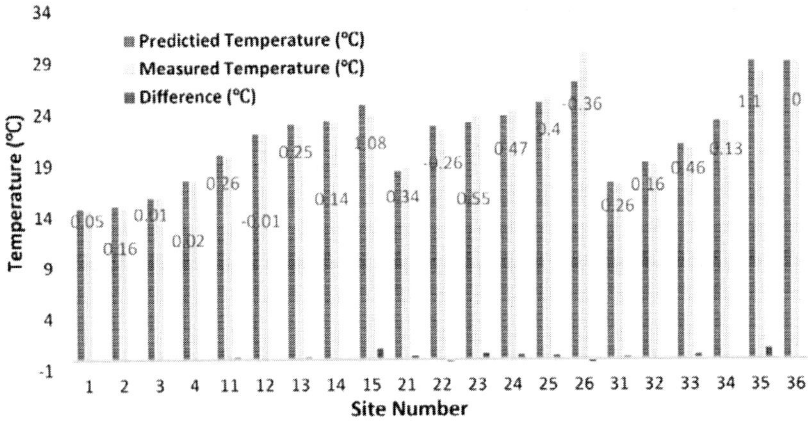


**Figure 4.** Comparison of predicted and field measured temperature.

Based on the case study, the proposed underground air-route temperature prediction model appears to represent the characteristics of temperature increase underground. Various underground heat sources (air compression, wall oxidation, underground heat, machinery, *etc.*) are covered by the model. By applying the law of heat conservation, a mathematical solution for deriving temperature was proposed via thermodynamic equations. A testing case for an ultra-deep underground coal mine demonstrated the accuracy of the prediction model. However, the following points should be improved in future research; (a) more types of underground heat sources should be covered and the related calculation equations should be derived. Considering the structure of the developed model, incorporating new heat sources and expanding the model capability are not difficult; (b) an in-depth analysis of moisture exchange in underground roadway walls should also be performed. The latent heat energy may change the air temperature due to such an exchange.

**Table 5.** Derived temperature prediction results from the computer program for route 1.

| Site Number | Site Name | Temperature at Starting Point (°C) | Temperature Prediction at End Point (°C) | Measured Temperature at End Point (°C) | D-Value (°C) |
|---|---|---|---|---|---|
| 1 | Surface | 14.6 | 14.85 | 14.8 | 0.05 |
| 2 | Mine shaft bottom | 14.8 | 15.06 | 14.9 | 0.16 |
| 3 | Airflow observation station | 14.9 | 15.91 | 15.9 | 0.01 |
| 4 | −700 North level airflow split point | 15.9 | 17.62 | 17.6 | 0.02 |
| 11 | $-700_{-4}$ level roadway | 17.6 | 20.06 | 19.8 | 0.26 |
| 12 | $-700_{-4}$ level roadway | 19.8 | 22.09 | 22.1 | −0.01 |
| 13 | 7407 air-intake entry | 22.1 | 23.05 | 22.8 | 0.25 |
| 14 | 7407 working face head | 22.8 | 23.34 | 23.2 | 0.14 |
| 15 | 7407 working face end | 23.2 | 24.88 | 23.8 | 1.08 |

**Table 6.** Derived temperature prediction results from the computer program for route 2.

| Site Number | Site Name | Temperature at Starting Point (°C) | Temperature Prediction at end Point (°C) | Measured Temperature at End Point (°C) | D-Value (°C) |
|---|---|---|---|---|---|
| 1 | Surface | 14.6 | 14.85 | 14.8 | 0.05 |
| 2 | Mine shaft bottom | 14.8 | 15.06 | 14.9 | 0.16 |
| 3 | Airflow observation station | 14.9 | 15.91 | 15.9 | 0.01 |
| 4 | −700 North level airflow split point | 15.9 | 16.21 | 16.6 | 0.39 |
| 21 | $-700_{-3}$ level roadway | 16.6 | 18.46 | 18.8 | 0.34 |
| 22 | $-700_{-3}$ level roadway | 18.8 | 22.86 | 22.6 | −0.26 |
| 23 | $-700_{-3}$ level roadway | 22.6 | 23.15 | 23.7 | 0.55 |
| 24 | 7301 air-intake entry | 23.7 | 23.83 | 24.3 | 0.47 |
| 25 | 7301 working face head | 24.3 | 25.1 | 25.5 | 0.4 |
| 26 | 7301 working face end | 26.1 | 27.08 | 29.8 | −0.36 |

**Table 7.** Derived temperature prediction results from the computer program for route 3.

| Site Number | Site Name | Temperature at Starting Point (°C) | Temperature Prediction at End Point (°C) | Measured Temperature at End Point (°C) | D-Value (°C) |
|---|---|---|---|---|---|
| 1 | Surface | 14.6 | 14.85 | 14.8 | 0.05 |
| 2 | Mine shaft bottom | 14.8 | 15.06 | 14.9 | 0.16 |
| 3 | Air measuring station | 14.9 | 15.91 | 15.9 | 0.01 |
| 31 | −700 South level airflow split point | 16.4 | 17.36 | 17.1 | 0.26 |
| 32 | −700 level roadway | 17.1 | 19.26 | 19.1 | 0.16 |
| 33 | −700 level roadway | 19.1 | 21.06 | 20.6 | 0.46 |
| 34 | Dip head entry in south area | 20.6 | 23.33 | 23.2 | 0.13 |
| 35 | Dip head entry in south area | 23.2 | 29.1 | 28 | 1.1 |
| 36 | Mine development head | 28 | 29 | 29 | 0 |

## ACKNOWLEDGMENTS

This work is financially supported by grants from the National Science Foundation of China (Grant No.51304203), the Natural Science Foundation of Jiangsu Province of China for Youths (Grant No. BK20130191) and Specialized Research Fund for the Doctoral Program of Higher Education (Grant No. 20130095120001); the authors are appreciating for these supports.

## AUTHOR CONTRIBUTIONS

Shuai Zhu and Shiyue Wu conceived and designed the mathematical model and field testing, Jianwei Cheng deals with the math calculation and analyzes the data and improve the paper quality, Siyuan Li improves the figure works and writes the part of introduction; Mingming Li collects the background materials.

## REFERENCES

1. SACMS (2007). State Administration of Coal Mine Safety. Available online: http://www.chinasafety.gov.cn/2007-02/28/content_ 220502.htm (assessed on 8 August 2015).
2. SACMS (2012). State Administration of Coal Mine Safety. Available online: http://www.china.com.cn/policy/txt/2011-12/07/content_24092977.htm (accessed on 8 August 2015).
3. He, M. Application of HEMS cooling technology in deep mine heat hazard control. *Min. Sci. Technol.***2009**, 3, 269–275.
4. Brake, R. Fluid Consumption, Sweat Rates and Hydration Status of Thermally Stressed Underground Miners and the Implications for Heat Illness and Shortened Shifts. Available online: http://www.qrc.org.au/conference/_dbase_upl/2001_spk024_brake.pdf (a ccessed on 4 August 2015).

5. United States Disbursing Officer Administration. *Heat Stress in Mining*; United States Disbursing Officer Administration: Washington, DC, USA, 2014.

6. Bossard, F. Calculating underground mine air-cooling requirements. In Proceedings of the 6th US Mine Ventilation Symposium, Salt Lake, UT, USA, 21–23 June 1993.

7. Torres, F.; Singh, R. Thermal State and Human Comfort in Underground Mining. In *Developments in Heat Transfer*; Aurelio, M., Bernardes, S., Eds.; INTECH Open Access Publisher: Rijeka, Croatia, 2011.

8. Jian, C.G.; Su, W.; Zhang, H.; Yang, F. Application of cold water spray in mine heat hazard control.*Appl. Mech. Mater.* **2011**, *71–78*, 2375–2381.

9. Wang, S.; Ren, T.; Zhang, T.; Liang, Y.; Xu, Z. Hot environment—Estimation of thermal comfort in deep underground mines. In Proceedings of the 12th Coal Operators' Conference, Wollongong, Australia, 16–17 February 2012; University of Wollongong & the Australasian Institute of Mining and Metallurgy: Wollongong, Australia, 2012; pp. 241–248.

10. Wagner, H. The management of heat flow in deep mines (Part 2). *Geomechan. Tunn.* **2011**, *4*, 157–163.

11. Hartman, H.L.; Mutmansky, J.M.; Ramani, R.V.; Wang, Y.J. *Mine Ventilation and Air Conditioning*; Wiley-Interscience: Hoboken, NJ, USA, 1997.

12. Payne, T.; Mitra, R. A review of heat issues in underground metalliferous mines. In *12th U.S./North American Mine Ventilation Symposium*; University of Nevada: Reno, NV, USA, 2008.

13. Qin, Y.; Hou, Q. *Environmental Engineering in Mines*; Wuhan Industrial University Press: Wuhan, China, 1989.

14. Du Plessis, J. *Ventilation and Occupational Environment Engineering in Mines*; Mine Ventilation Society of South Africa: Johannesburg, South Africa, 2014.

15. Yang, D.; Yang, T. *Thermal Environments in Mine and Its Control*; Metallurgical Industry Press: Beijing, China, 2009.

# CHAPTER 5

# Analysis on the Initial Cracking Parameters of Cross-Measure Hydraulic Fracture in Underground Coal Mines

**Yiyu Lu [1,2], Liang Cheng [1,*], Zhaolong Ge [1,2,*], Binwei Xia [1,2], Qian Li [1,2] and Jiufu Chen [3]**

[1]*State Key Laboratory of Coal Mine Disaster Dynamics and Control, Chongqing University, Chongqing 400044, China*

[2]*National & Local Joint Engineering Laboratory of Gas Drainage in Complex Coal Seam, Chongqing University, Chongqing 400044, China*

[3]*Songzao Coal-Electricity Limited Liability Company, Chongqing 401445, China*

## ABSTRACT

Initial cracking pressure and locations are important parameters in conducting cross-measure hydraulic fracturing to enhance coal seam permeability in underground coalmines, which are significantly influenced by *in-situ* stress and occurrence of coal seam. In this study, stress state around cross-measure fracturing boreholes was analyzed using *in-situ* stress coordinate transformation, then a mathematical model was developed to evaluate initial cracking parameters of borehole assuming the maximum tensile stress criterion. Subsequently, the influences of *in-situ* stress and occurrence of coal seams on initial cracking pressure and locations in underground coalmines were analyzed using the proposed model. Finally, the proposed model was verified with field test data. The results suggest that the initial cracking pressure increases with the depth cover and coal seam dip angle. However, it decreases with the increase in azimuth of major principle stress. The results also indicate that the initial cracking locations concentrated in the second and fourth quadrant in polar coordinate, and shifted direction to the strike of coal seam as coal seam dip angle and azimuth of maximum principle stress increase. Field investigation revealed consistent rule with the developed model that the initial cracking pressure increases with the coal seam dip angle. Therefore, the proposed mathematical model provides theoretical insight to analyze the initial cracking parameters during cross-measure hydraulic fracturing for underground coalmines.

## KEYWORDS

coal seam; hydraulic fracturing; *in-situ* stress; coal seam dip angle; initial cracking parameter

## 1. INTRODUCTION

High efficient extraction of coal bed methane is of great importance to the safety of coalmine production and the development of state energy strategy in China [1]. The gas pre-extraction technique by boreholes has been widely used to recover the underground coal bed methane and hence prevent coalmine disasters [2,3]. In practice, the permeability of coal seams plays an important role in coal bed methane extraction [4,5,6]. Research efforts in recent year have proved that the hydraulic fracturing technology can enhance the coal seam permeability significantly, and has been used during the stages of coal roadway development and rock cross-cut coal uncovering, where a special technique of cross-measure hydraulic fracture is usually used [7]. In the implementation of cross-measure hydraulic fracturing, boreholes are drilled through seam floor into the targeted coal seam in a special gas roadway or main gate road. The typical spatial relationship between the borehole and coal seams is shown in Figure 1. In mine sites, pumping pressure of hydraulic fracturing is empirically elevated to seek a wide range of hydraulic fracture. However, as a side effect, meanwhile this incurs the risk of damaging the integrity of coal seam roof and floor, making the stability of roadway difficult to maintain. Therefore, it is very important to analyze the initial cracking parameters of cross-measure hydraulic fracturing. This would help reveal the initiation mechanism and provide theoretical guidance for implementing suitable cross-measure hydraulic fracturing technology.

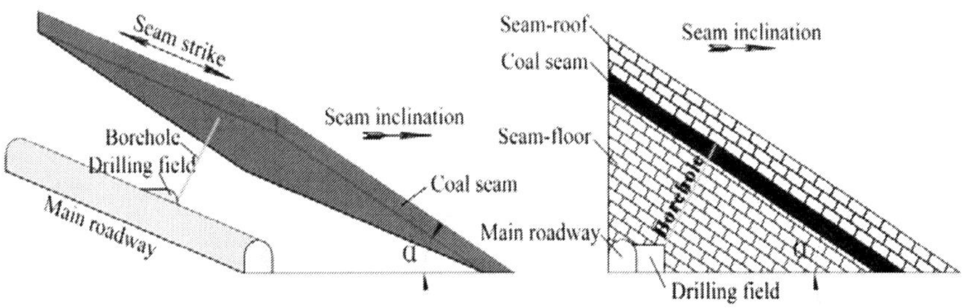


**Figure 1.** Relationship between the fracturing boreholes and coal seams.

In order to better understand the fracturing initiation mechanism, many studies have been performed in the literature, and substantial studies have been conducted focusing on the influences of the *in-situ* stress, coal and rock mass strength and pore pressure. For the first time, Hubbert and Willis reported the development of longitudinal cracks in vertical boreholes in oil and gas well fracturing activities, and established the correlation of the horizontal principle stress, rock tensile strength, and pore pressure with pump pressure [8]. Schmidt and Zoback considered the influence of porosity and Poisson ratio on the pressure of crack initiation when studying the initiation mechanism in both permeable and impermeable rocks [9]. Zhang and Jeffrey investigated the effect of major and intermediate principle stresses on the crack initiation and propagation, and discussed the influence of misalignment and crack numbers on the crack initiation location near the well [10]. Huang and Griffiths *et al.* found the relationship between the initiation pressure, initiation location, azimuth of cracks and deviation well angles, and well azimuth by transforming the coordinates of *in-situ* stress to well wall stress [11,12]. Chen *et al.* found that borehole wall crack initiation was influenced by the level and direction of *in-situ* stress as well as borehole dig angle [13]. Using two-dimensional particle flow code—$PFC^{2D}$, Wang *et al.* investigated relevant parameters and analyzed the correlation between macroscopic and mesoscopic mechanical parameters [14]. Zhang *et al.* established a two-dimensional numerical fracture model to analyze the coupled rock deformation and fluid flow in such fractures, and the results show that fracture escape from the interface is likely under the conditions of fracture growth from stiff to soft rocks, small layer-to-layer far-field stress contrasts, and moderately low fluid viscosity, and small parent fracture lengths and offset distances [15,16].

Although substantial studies have been conducted advancing the technology of hydraulic fracture, it should be noted that most studies are performed from the viewpoint of oil and gas industry. Different from hydraulic fracturing in oil and gas extraction, hydraulic fracturing in underground coalmines are of the following salient features:

(1)Hydraulic fracturing in underground coalmines is an auxiliary technology, serving subsequent coal mining. Therefore, monotonic increase of the pump pressure during the fracturing process is not allowed to avoid damaging the integrity of coal seam roof and floor. In this aspect, Lin studied the pulsating stress wave generation, propagation and the mechanism of coal and rock breakage from the theoretical perspective [17].

(2)The coal bed is thin and bounded by coal seam roof and floor, and it is conventionally assumed that the physical and mechanical properties of coal would not affect hydraulic fracturing initiation. However, hydraulic fracturing initiation can be influenced by other factors, such as *in-situ* stress and rock strength. Using numerical simulation, Lian and Lin, *et al.* found the rule of crack initiation and propagation influenced by *in-situ* stress and rock strength [18,19].

(3)The layout of a borehole is limited by the occurrence of coal seam such as strike, inclination, and dip angle as shown in Figure 1.

Therefore, the crack initiation and propagation of coalmines by hydraulic fracturing is directly influenced by coal seam occurrence, especially by factors of strike, inclination and dip angle and *in-situ* stress of coal seam, and cannot be interpreted with well-developed hydraulic fracturing theory from the oil and gas industry.

The purpose of this study is to investigate the influences of the *in-situ* stress and coal seam dip angle on the initial cracking pressure and locations in hydraulic fracturing of underground coalmines. In this study, the stress state around cross-measure fracturing boreholes was analyzed using *in situ* stress coordinate conversion, and then a model to calculate the initiation parameters of the borehole was developed using the maximum tensile stress criterion. Finally, the developed mathematical model was verified with field data.

# 2. MODEL DEVELOPMENT

## 2.1. Stress Analysis for Cross-Measure Fracturing Boreholes

The stress state of a coal seam is shown in Figure 2. The analysis of the stress state around borehole was firstly performed based on the following assumptions:

(1) Borehole is perpendicular to coal seam, as shown in Figure 1.

(2) Taking the direction of coal seam inclination as 0°, corresponding to 0-$x_1$ in Figure 2.

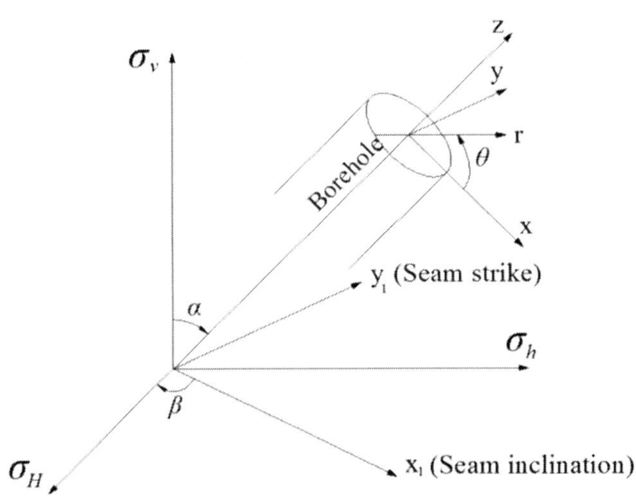


**Figure 2.** Stress state of the borehole.

The stress state on borehole wall can be obtained by converting the coordinate of *in-situ* rock stress [12].

$$\begin{cases} \sigma_x = \left(\sigma_H \cos^2\beta + \sigma_h \sin^2\beta\right)\cos^2\alpha + \sigma_v \sin^2\alpha \\ \sigma_y = \sigma_H \sin^2\beta + \sigma_h \cos^2\beta \\ \sigma_z = \left(\sigma_H \cos^2\beta + \sigma_h \sin^2\beta\right)\sin^2\alpha + \sigma_v \cos^2\alpha \\ \tau_{xy} = \left(\sigma_h - \sigma_H\right)\cos\alpha\cos\beta\sin\beta \\ \tau_{yz} = \left(\sigma_h - \sigma_H\right)\sin\alpha\cos\beta\sin\beta \\ \tau_{zx} = \dfrac{\left(\sigma_H \cos^2\beta + \sigma_h \sin^2\beta - \sigma_v\right)}{2}\sin 2\alpha \end{cases} \tag{1}$$

where $\sigma_v$ is the vertical principal stress (MPa); $\sigma_H$ and $\sigma_h$ are the major and minor horizontal principal stress (MPa); α is the coal seam dip angle (°); and β is the azimuth of the major principle stress (°).

Then, stress components at a distance of $r$ from borehole caused by principle stress $\sigma_H$, $\sigma_h$, and $\sigma_v$ can be obtained from Equation (1) [12]:

$$\begin{cases} \sigma_r = \dfrac{R^2}{r^2}p + \dfrac{\sigma_x+\sigma_y}{2}\left(1-\dfrac{R^2}{r^2}\right) + \left(1+\dfrac{3R^4}{r^4}-\dfrac{4R^2}{r^2}\right)\dfrac{\sigma_x-\sigma_y}{2}\cos 2\theta + \tau_{xy}\left(1+\dfrac{3R^4}{r^4}-\dfrac{4R^2}{r^2}\right)\sin 2\theta \\ \sigma_\theta = \dfrac{\sigma_x+\sigma_y}{2}\left(1+\dfrac{R^2}{r^2}\right) - \dfrac{R^2}{r^2}p - \dfrac{\sigma_x-\sigma_y}{2}\left(1+\dfrac{3R^4}{r^4}\right)\cos 2\theta - \tau_{xy}\left(1+\dfrac{3R^4}{r^4}\right)\sin 2\theta \\ \sigma_{zz} = \sigma_z - \nu[2\left(\sigma_x+\sigma_y\right)\dfrac{R^2}{r^2}\cos 2\theta + 4\tau_{xy}\dfrac{R^2}{r^2}\sin 2\theta] \\ \tau_{r\theta} = \dfrac{\sigma_y-\sigma_x}{2}\left(1-\dfrac{3R^4}{r^4}+\dfrac{2R^2}{r^2}\right)\sin 2\theta + \tau_{xy}\left(1-\dfrac{3R^4}{r^4}+\dfrac{2R^2}{r^2}\right)\cos 2\theta \\ \tau_{\theta z} = \tau_{yz}\left(1+\dfrac{R^2}{r^2}\right)\cos\theta - \tau_{xz}\left(1+\dfrac{R^2}{r^2}\right)\sin\theta \\ \tau_{zr} = \tau_{xz}\left(1-\dfrac{R^2}{r^2}\right)\cos\theta + \tau_{yz}\left(1-\dfrac{R^2}{r^2}\right)\sin\theta \end{cases} \tag{2}$$

where $R$ is the radius of the borehole, $p$ is the water pressure in the borehole, and $\sigma_r$, $\sigma_\theta$, $\sigma_{zz}$, $\tau_{r\theta}$, $\tau_{\theta z}$, and $\tau_{zr}$radial radial, tangential and axial components of normal and shear stress (MPa) at a distance of $r$ from borehole and inclined at an angle of θ with $\sigma_y$, respectively.

## 2.2. Model for Calculation of Initial Cracking Parameters

To simplify the analysis, the implementation of cross-measure fracturing is assumed to be conducted in an elastomeric tube. In other words, the derivation only considers the factors of high water pressure and *in-situ*stress. The stress state of borehole walls is given by Equation (2), where *r* is defined as *R*:

$$\begin{cases} \sigma_r = p \\ \sigma_\theta = (\sigma_x + \sigma_y) - 2(\sigma_x - \sigma_y)\cos 2\theta - 4\tau_{xy}\sin 2\theta - p \\ \sigma_{zz} = \sigma_z - \nu\left[2\left(\sigma_x + \sigma_y\right)\cos 2\theta + 4\tau_{xy}\sin 2\theta\right] \\ \tau_{\theta z} = 2\tau_{yz}\cos\theta - 2\tau_{xz}\sin\theta \\ \tau_{r\theta} = \tau_{zr} = 0 \end{cases} \quad (3)$$

The principle stresses on borehole walls can be obtained using Equation (3):

$$\begin{cases} \sigma_1 = \sigma_r \\ \sigma_2 = \dfrac{1}{2}\left[\sigma_\theta + \sigma_{zz} + \sqrt{(\sigma_\theta - \sigma_{zz})^2 + 4\tau_{\theta z}{}^2}\right] \\ \sigma_3 = \dfrac{1}{2}\left[\sigma_\theta + \sigma_{zz} - \sqrt{(\sigma_\theta - \sigma_{zz})^2 + 4\tau_{\theta z}{}^2}\right] \end{cases} \quad (4)$$

From Equation (4), it can be seen that the initial cracking would occur in θ–*z* plane, as shown in Figure 3. According to the maximum tensile stress criterion, the crack initiation on borehole walls would occur when the tensile stress meets the ultimate tensile strength of coal [20]. Using Equation (4), the maximum tensile stress on borehole wall can be obtained by:

$$\sigma_{max} = \sigma_3 = \frac{1}{2}\left[\sigma_\theta + \sigma_{zz} - \sqrt{(\sigma_\theta - \sigma_{zz})^2 + 4\tau_{\theta z}^{\ 2}}\right] \tag{5}$$

Using the concept of effective stress, the stress contributed to initiate fracture (termed "fracturing stress" in this paper) on the borehole wall can be estimated as:

$$\sigma_f = \sigma_{max} - p_0 \tag{6}$$

where p0 is pore pressure (MPa).

The fracturing stress $\sigma_f$ is a function of borehole pressure $p$ through $\sigma_3$, which is a function of $\sigma\theta$ directly related to $p$. However, the value of $\sigma_f$ is not a constant at angular positions around the borehole wall. In hydraulic fracturing process, the fracture starts to grow at an angular position, say $\theta$, with the increase of borehole pressure $p$ from zero to the threshold. The condition for borehole wall cracking initiation can be given by Hossain *et al.* [21]:

$$\sigma_f \geq -\sigma_t \tag{7}$$

where $\sigma_t$ is the coal tensile strength (MPa).

The initial angular position of cracks is obtained by:

$$\begin{cases} \dfrac{\partial \sigma_f}{\partial \theta} = 0 \\ \dfrac{\partial^2 \sigma_f}{\partial \theta^2} > 0 \end{cases} \tag{8}$$

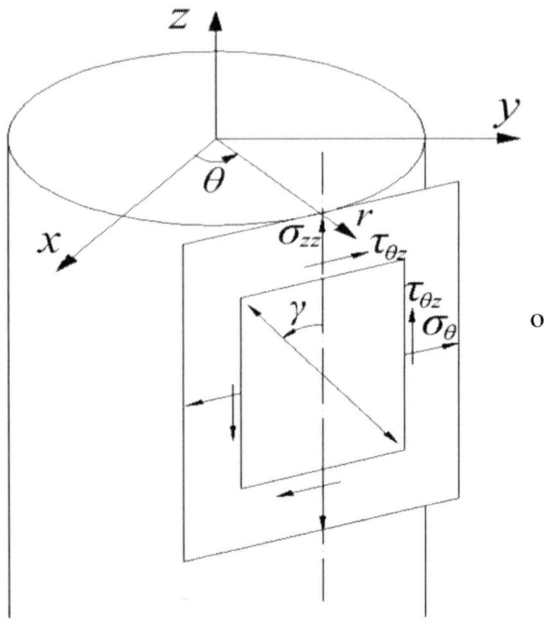


**Figure 3.** Stress state on the borehole wall.

When $\theta = \theta_f$, the minimum value of $\sigma_f$ in Equation (6) can be found. If this value meets the requirement of Equation (7), the borehole will initiate. So, based on Equations (6) and (7), the fracture initiation criterion can be obtained:

$$\sigma_{\max} - p_0 + \sigma_t = 0 \tag{9}$$

From Equation (9), it can be seen that rock mass of high tensile strength requires a high fracture initiation pressure, while the pore pressure $p_0$ has a positive effect on hydraulic fracture initiation by reducing the required pressure. The research shows that the pore pressure and tensile strength of rock can be assumed as zero to achieve a conservative pressure for fracture initiation [22,23,24]. Therefore, the solution of fracture initiation criterion (Equation (9)) can be simplified as:

$$\sigma_{\max} = 0 \tag{10}$$

Equations (5) and (10) can be combined to obtain the commonly used relationship

$$\begin{aligned} &\frac{1}{2}\left[\sigma_{\theta} + \sigma_{zz} - \sqrt{(\sigma_{\theta} - \sigma_{zz})^2 + 4\tau_{\theta z}^{\ 2}}\right] = 0 \\ &\sigma_{\theta}\sigma_{zz} - \tau_{\theta z}^{\ 2} = 0 \end{aligned} \tag{11}$$

Equations (3) and (11) can be combined to obtain the simplified fracture initiation criterion

$$\begin{aligned} &\left[(\sigma_x + \sigma_y) - 2(\sigma_x - \sigma_y)\cos 2\theta_f - 4\tau_{xy}\sin 2\theta_f - p\right]\sigma_{zz} - \tau_{\theta z}^{\ 2} = 0 \\ &p_f = (\sigma_x + \sigma_y) - 2(\sigma_x - \sigma_y)\cos 2\theta_f - 4\tau_{xy}\sin 2\theta_f - \frac{\tau_{\theta z}^{\ 2}}{\sigma_{zz}} \end{aligned} \tag{12}$$

As can be seen from Equations (1), (3) and (12), the initial cracking pressure ($p_f$) and location ($\theta_f$) of cross-measure fracturing are affected conjointly by coal seam dip angle (α), azimuth of maximum principle stress (β), and *in-situ* stress.

## 3. FIELD EXPERIMENTATION

Based on the developed model of crack initiation, hydraulic fracturing trials were carried out at Tonghua coalmine in Songzao mining area (Chongqing, China), and the relationship between the crack initiation pressure and the coal seam dip angle is verified with field data. The coal seam inclination is assumed to be 0° during the test.

### 3.1. Mine-Site Condition and Borehole Layout

The hydraulic fracturing trials are conducted at ±0 m south main roadway of Tonghua coalmine, and the purpose is to drain the coal bed methane from the No. 3121 driving band air return roadway efficiently, and then tunnel the No. 3121 driving band air return roadway. There are three coal seams in this area. $K_1$ coal seam is the targeted seam for fracturing that is usually called the "floor chamber", with a black or slightly steel-gray color, which is a single seam that has a layered structure and massive textures with luster of metallic or semi-metallic. The depth of $K_1$ coal seam is 500 m, with an average thickness of 1.65 m, dip angle ranging from 29.88° to 48.00°. Eight boreholes are drilled along strike of No. 3121 driving band air return roadway. Profile map of fracturing drilling is illustrated in Figure 4 and drilling parameters are listed inTable 1. The detailed sealing material and sealing craft in fracturing were used referring to Ge *et al.* [25].

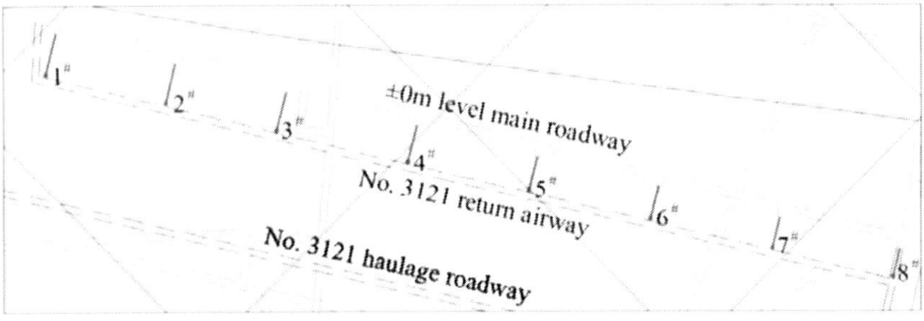


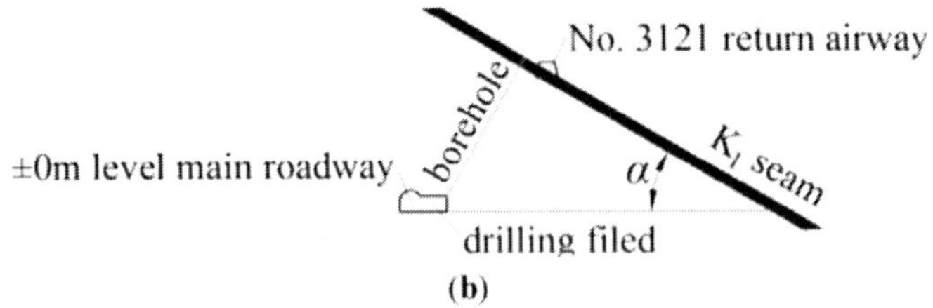


**Figure 4.** The layout of the fracturing boreholes: (**a**) the plan of the fracturing boreholes; (**b**) the section of the fracturing boreholes.

**Table 1.** Fracturing boreholes parameters.

| Number | Borehole Size/mm | Azimuth Angle/° | Borehole Angle/° | Borehole Depth/m |
|---|---|---|---|---|
| 1# | 91 | 0 | 60 | 30.2 |
| 2# | 91 | 0 | 57 | 33.2 |
| 3# | 91 | 0 | 55 | 35.1 |
| 4# | 91 | 0 | 53 | 37.0 |
| 5# | 91 | 0 | 50 | 39.8 |
| 6# | 91 | 0 | 47 | 42.4 |
| 7# | 91 | 0 | 44 | 45.0 |
| 8# | 91 | 0 | 42 | 46.6 |

## 3.2. In-Situ Stress Measurement in Test Site

In order to accurately obtain the relationship between the initial cracking pressure and coal seam dip angle, the technology of Kaiser effect of rock acoustic emission was used to measure *in-situ* stress of the rocks in fractured area. Rock cores were collected from six special directions in the measuring points of fractured rock, based on the fact that the rock has a characteristic of memorizing the original stress level. Subsequently, the rock core specimens were tested using the uniaxial compression apparatus. Kaiser points were determined from the test results, and the stress level were derived from Kaiser points, then the level and directions of *in-situ* stress at measuring points can be calculated using the theory of elasticity [26,27].Figure 5 shows the rock core layout for *in-situ* stress tests at ±0 m south main roadway of Tonghua coalmine. Note that the convention "$x$" in Figure 5 is equal to "$x_1$" in Figure 2.

The specimens of a length-to-width ratio 2:1 were cored in the field and three samples were collected in each direction (Figure 6). The main equipment for experiment includes electronic precision material testing machine and acoustic emission detection system. Table 2 shows the basic parameters of the equipment. In the loading stage, a velocity of displacement control was applied by 0.005 mm/s, and the change of ring-down count rate, energy count rate, accumulative ring-down count, accumulative energy count with elapse of time was recorded.

Then, according to the observed Kaiser effect, the *in-situ* stress environment was derived for test site (Table 3).

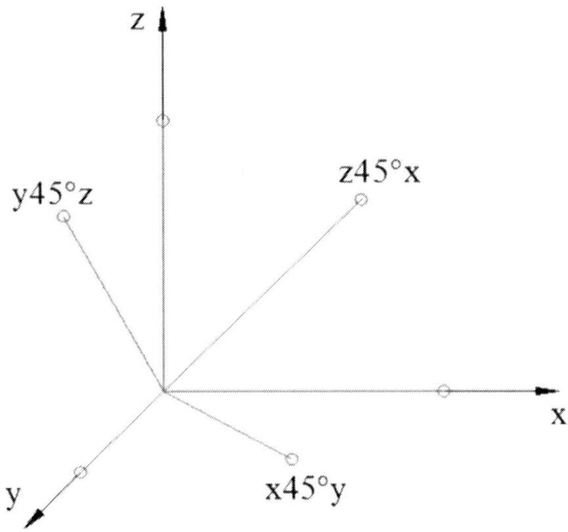


**Figure 5.** Boreholes layout for the rock core.

**Figure 6.** Specimens in acoustic emission test.

**Table 2.** Basic parameter of equipment.

| Equipment Name | Type | Parameters |
|---|---|---|
| Electronic precision material testing machine | AG-I 250kN | 1/1000 degree loading precision, 1/1000 degree displacement control precision |
| Acoustic emission detection system | 16CHsSAMOS™ System | 12 channels, 40 MHz, 18 bit A/D, 1 KHz–3 MHz frequency range, 100 dB maximum signal amplitude, 40 MSPS sample rate |

**Table 3.** Test result of *in-situ* stress.

| Category | The Maximum $\sigma_1$ | The Intermediate $\sigma_2$ | The Minimum $\sigma_3$ |
|---|---|---|---|
| Principal stress /MPa | 23.7 | 9.9 | 4.5 |
| $\theta_x$ (The angle between principal stress and $x$)/° | 39.0 | 94.6 | 51.1 |
| $\theta_y$ (The angle between principal stress and $y$)/° | 129.1 | 87.8 | 38.9 |
| $\theta_z$ (The angle between principal stress and $z$)/° | 86.8 | 3.4 | 88.4 |

## 3.3. Hydraulic Fracturing Equipment for Field Test

Hydraulic fracturing test was conducted using two fracturing pumps (Liuhe Coal Mining Machinery Co. Ltd, Nanjing, China) at a flow rate of 200 L/min and a maximum working pressure of 56 MPa. As the pumps are variable displacement pumps, the pumps are used in pressure control mode. The high-pressure hydraulic fracturing system mainly consists of the fracturing pump, consoles, water tanks, water meters, pressure gauges, high-pressure pipes, sealing accessories, and relevant connecting devices, as shown inFigure 7. The fracturing fluid is water. During the fracturing process, two pumps run simultaneously. Figure 8 shows the field test of the cross-measure hydraulic fracturing in underground coalmines.

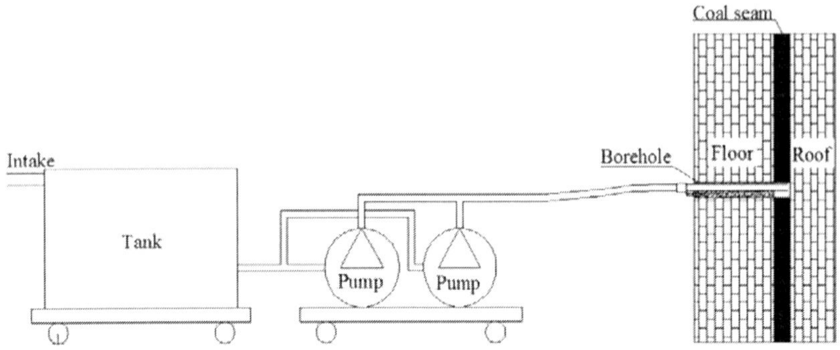


**Figure 7.** The connected equipment of the high-pressure hydraulic fracturing system.

**Figure 8.** Field test of cross-measure hydraulic fracturing in underground coalmines.

# 4. RESULTS AND DISCUSSION

## 4.1. Analysis on the Initiation Parameters

### 4.1.1. State of *In-Situ* Stress in Coal Seams

Underground, a coal seam is in a triaxial compressive stress state, where three principle stress components are usually unequal to each other. The vertical stress component $\sigma_v$ is equal to the overlying rock weight in vertical orientation. There is a linear relationship between the major or intermediate principle stress and the seam burial depth. The *in-situ* stress analysis is assumed to obey the stress distribution rule that is suitable for Mainland China [28,29]. Thus, the principle stresses in Equation (1) can be obtained:

$$\begin{cases} \sigma_v = \rho H \\ \sigma_H = 6.7808 + 0.0216H \\ \sigma_h = 2.2323 + 0.0182H \end{cases} \tag{13}$$

where $\rho$ is the average weight of overlying rock (kN/m$^3$) and $H$ is the burial depth of coal seam (m). The burial depth of Chinese coal seam usually ranges from 300 to 700 m. The overlying rock is mainly composed of mudstone and marlstone strata. Conventionally, the average weight of overlying strata is taken as 25 kN/m$^3$. For simplicity, the Poisson ratio can be assumed as zero [13,21].

### 4.1.2. The Effect of Burial Depth on Borehole Cracking Initiation

In China, a coal seam is usually at a depth of less than 1000 m. The initial cracking pressure and location are plotted in Figure 9, where the values of $\alpha$ and $\beta$ are set as 0°, 30°, 45°, 60° or 90°. It can be seen that the initiation pressure increases with the buried depth of coal seam, and boreholes are more likely to collapse at a smaller initiation pressure under the *in-situ* stress. Figure 10 shows the initiation direction is independent of the burial depth, and the cracks initiated along the inclination or strike if α and β are both at 0° or 90°. When α and β are both at 30°, 45° or 60°, the initiation locations are mainly distributed in the second or fourth quadrant of the polar coordinate, but it tends to shift their direction to coal seam inclination with the increase of burial depth.

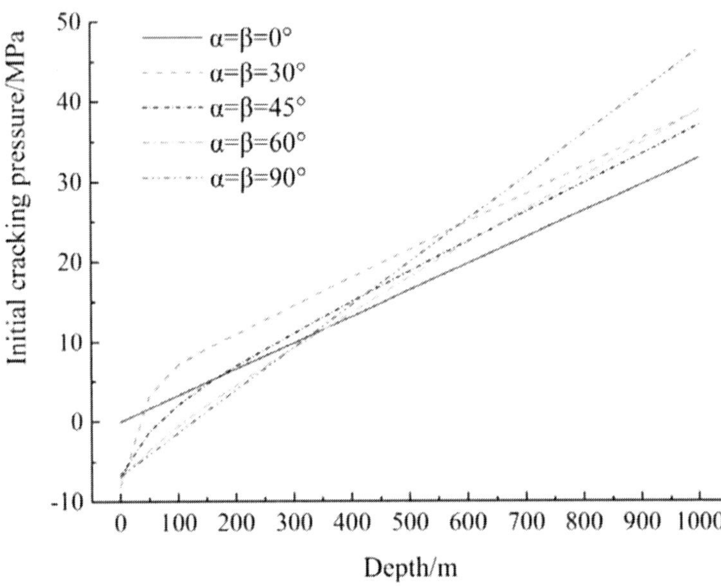


**Figure 9.** Relationship between the initial cracking pressure and burial depth.

### 4.1.3. The Effect of the Coal Seam Dip Angle on Borehole Cracking Initiation

For coal seam at a depth of 500 m, its stress state can be calculated using Equation (10), where $\sigma_v$, $\sigma_H$, $\sigma_h$ are 12.5, 17.58 and 11.33 MPa, respectively. Thus, when β is set at 0°, 30°, 45°, 60° and 90°, for different coal seam dip angle, the initiation pressure and initiation location changes as follows: the initiation pressure increases with coal seam dip angle as shown in Figure 11, and the cracks initiate along the inclination and strike of coal seam when β is set as 0° and 90°, respectively, as demonstrated in Figure 12. This trend is the same as that obtained with the increase in coal seam dip angle; when β is set as 30°, 45° or 60°, the cracks initiation locations are mainly distributed in the second or fourth quadrant of the polar coordinate. However, with the increase of coal seam dip angle, they shifted to the direction of coal seam strike, and the deflection speed increased during this process.

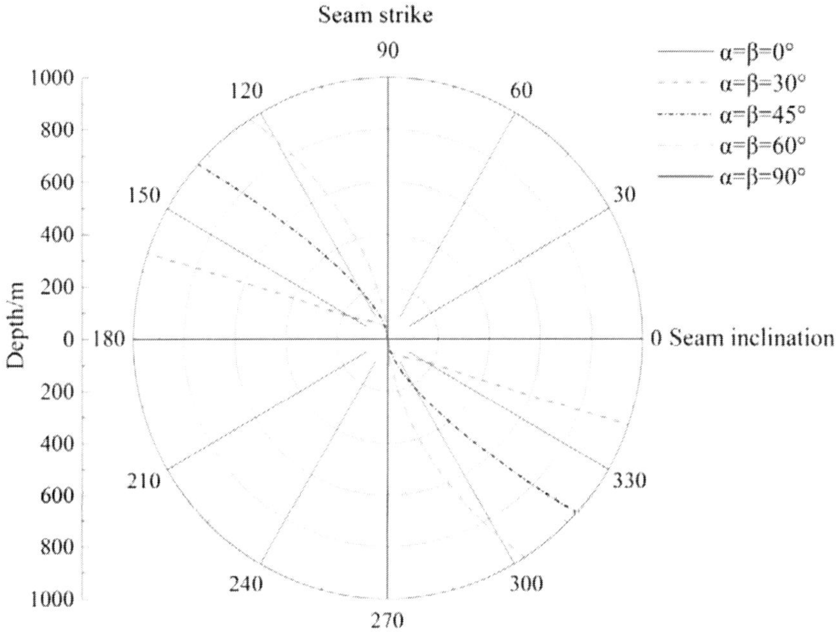


**Figure 10.** Relationship between the initial cracking location and burial depth.

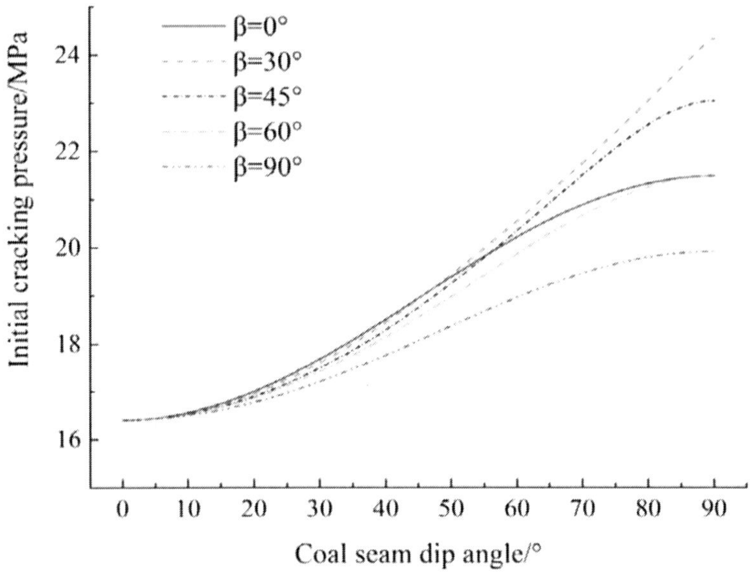


**Figure 11.** Relationship between the initial cracking pressure and coal seam dip angle.

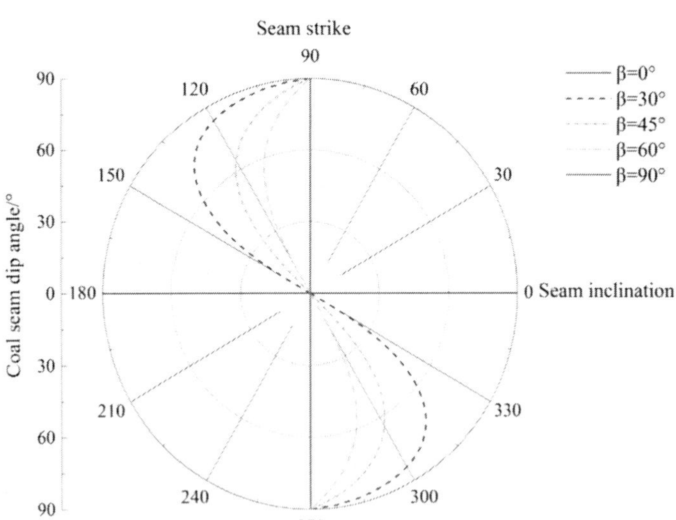


**Figure 12.** Relationship between the initial cracking location and coal seam dip angle.

#### 4.1.4. The Effect of the Stress Orientations on Boreholes Initiation

In this case, the same condition that the coal seam is buried at 500 m in depth is considered. When α is taken as 0°, 30°, 45°, 60° or 90°, for the different major principle stress azimuth, the initiation pressure and initiation location changes as follows.

As shown in Figure 13, when α is 0°, the initiation pressure remains invariable during the increase in the major principle stress azimuth. However, when α is taking the value of 30°, 45°, 60° and 90°, the initiation pressure decreases as the azimuth of major principle stress increases. It can be seen from Figure 14 that the cracks initiate along the direction of coal seam strike when α is 90°; when α is 0°, 30°, 45° and 60°, respectively, the initial cracking locations mainly occurred in the second or fourth quadrant of the polar coordinate, and shifted to direction of coal seam strike with the increase in azimuth angle. The deflection speed reduced in the process.

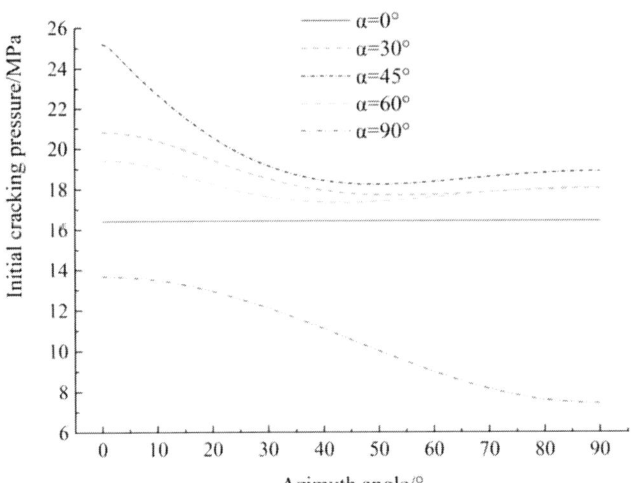


**Figure 13.** Relationship between the initial cracking pressure and the major principle stress azimuth angle.

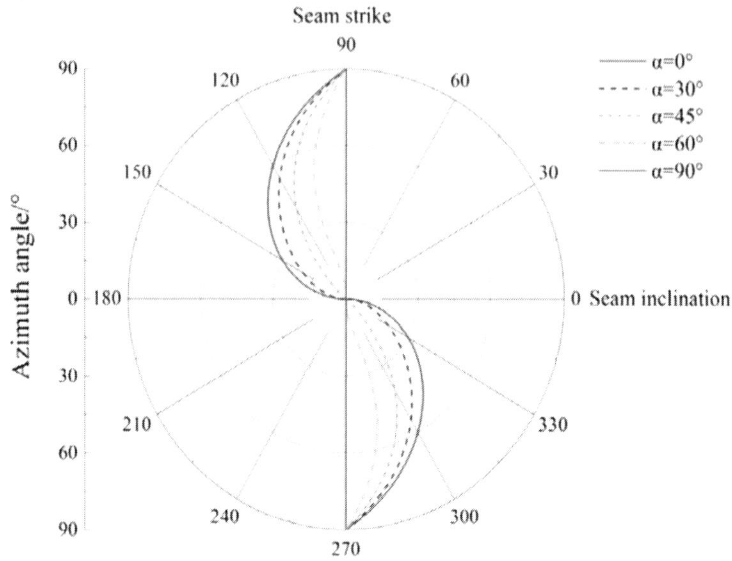


**Figure 14.** Relationship between the initial cracking location and the major principle stress azimuth angle.

## 4.2. Field Validation and Discussion

Table 3 summarizes the test results of *in-situ* stress at field site. The intermediate principal stress $\sigma_2$ and the vertical stress $\sigma_v$ are considered to be equal for the small offset of $\sigma_v$ in vertical direction. Due to the small offset between $\sigma_H$ and $\sigma_h$ in the horizontal plane, the major principal stress $\sigma_1$ and minor principal stress $\sigma_3$ are considered equal to the major horizontal principal stress $\sigma_H$ and minor horizontal principal stress $\sigma_h$, respectively. Using *in-situ* stress test results and the developed mathematical model for initial cracking parameters and the condition of coal seams in fractured area, the theoretical initiation pressure of boreholes from No. 1# to No. 8# were calculated individually and summarized in Table 4.

**Table 4.** Theoretical initiation parameters of Nos. 1#–8# borehole.

| Number | 1# | 2# | 3# | 4# | 5# | 6# | 7# | 8# |
|---|---|---|---|---|---|---|---|---|
| Initiation pressure/MPa | 16.8 | 17.3 | 17.9 | 18.4 | 19.7 | 20.8 | 21.9 | 22.6 |
| Initiation location/° | 144 | 143.5 | 143.1 | 142.7 | 142 | 141.2 | 140.6 | 139.7 |

Boreholes drilling and sealing were completed in Tonghua coalmine on 13–28 May 2013. From May 2013 to August 2013, the hydraulic fracturing were carried out in boreholes from No. 1# to No. 8# seriatim at ±0 m south main roadway of Tonghua coalmine. Figure 15 illustrates the pump pressure change in No. 1#borehole, during the fracturing process. Figure 16 shows the comparison between the theoretic initial cracking pressure and the initial cracking pump pressures. The points on the blue line in Figure 16 represent the initial cracking pump pressures of No. 1# to No. 8# borehole, respectively.

It can be seen from Figure 15 that the system pressure increased sharply at the beginning of the fracturing process, and then reduced when the cracks initiate. The decrease process of pump pressure is not the same as that during the oil and gas wells fracturing, but the repeated increase and decrease in the pressure can be observed. This can be explained by the reason that coal has a higher plasticity and more pores and cracks compared with rocks typically encountered during oil and gas well fracturing. Thus, the coal seam initial cracking pressure should be the first maximum pump pressure during fracturing. Then, the pump pressure of crack initiation for No. 1# to No. 8# borehole is shown in Figure 16. It can be seen that the pressure tends to increase with the coal seam dip angle, which is consistent with the theoretical prediction. However, the pump pressure of crack initiation is higher than that theoretical one. There maybe two reasons to explain this phenomenon:

(1) The fracturing pump is far away from the borehole for the safety purpose and a certain amount of pressure is dissipated in the pipe line

from the pump to borehole destination. This amount of pressure loss can be calculated using the following Equation [30]:

$$\Delta p_1 = \frac{59.7Q^2}{D^5\,\mathrm{Re}^{0.25}} L \qquad (14)$$

where Δp1 is the pressure loss of the pipe line (MPa/m); *Q* is the average flow (L/min); *D* is the diameter of pipe line (mm), which is 32 mm; *Re* is the Reynolds number, taking 11165*Q*/*D*; and *L* is the distance between pump and fractured borehole (m), which is usually set as 500 m.

(2) The gravity of water increases the actual initiation pressure due to the height difference between fractured borehole bottom and the discharged pump. This amount of pressure loss can be calculated from the following Equation:

$$\Delta p_2 = \rho g \Delta h \qquad (15)$$

where Δp2 is the pressure loss caused by the gravity of water (MPa); ρ is the density of water (kg/m3); *g* is the gravitational acceleration (m/s2); and Δh is the height difference (m), which can be calculated from Table 1.

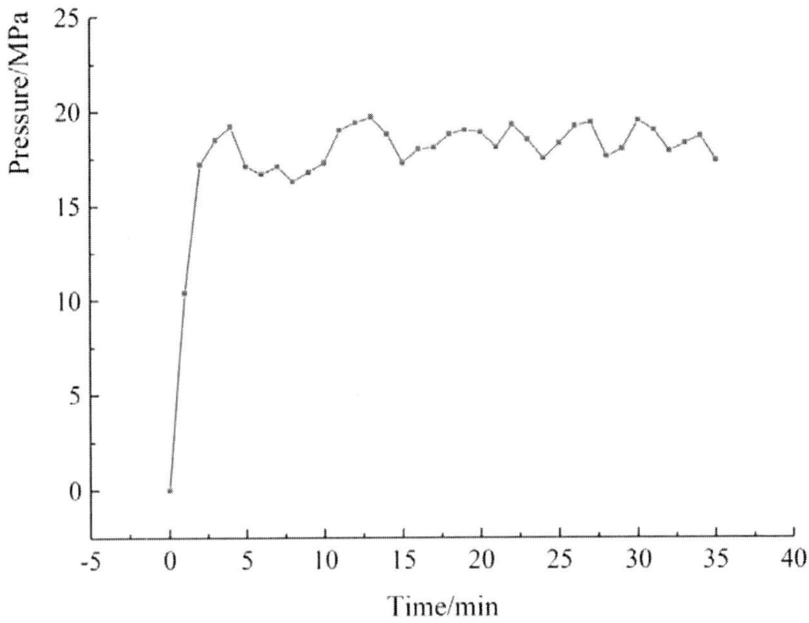


**Figure 15.** Pressure change in the borehole.

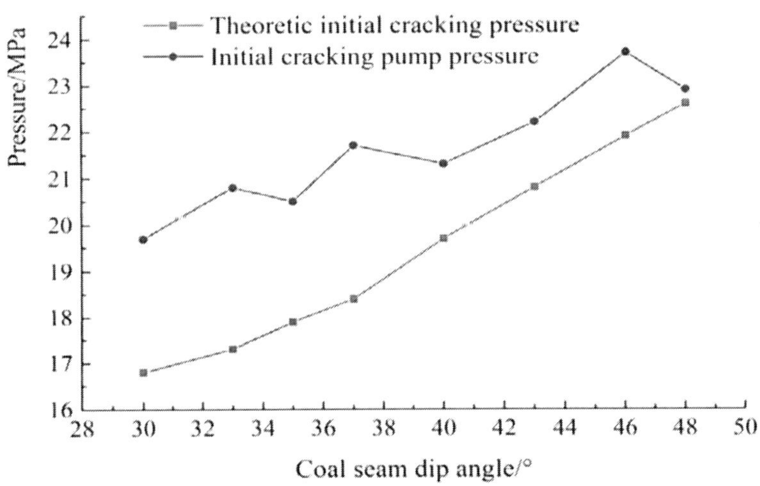


**Figure 16.** Comparison between the theoretic initial cracking pressure and the initial cracking pump pressure.

Therefore, after deducting the pressure loss of the pipeline and the water weight, the actual initiation pressure is shown in Figure 17. The actual initial cracking pressure is lower than the theoretical one, the natural joints and fractures of coal seam may decrease the initiation pressure [31], and the properties of the rock and the fracturing fluid can also affect the pressure to initiate hydraulic fracture.

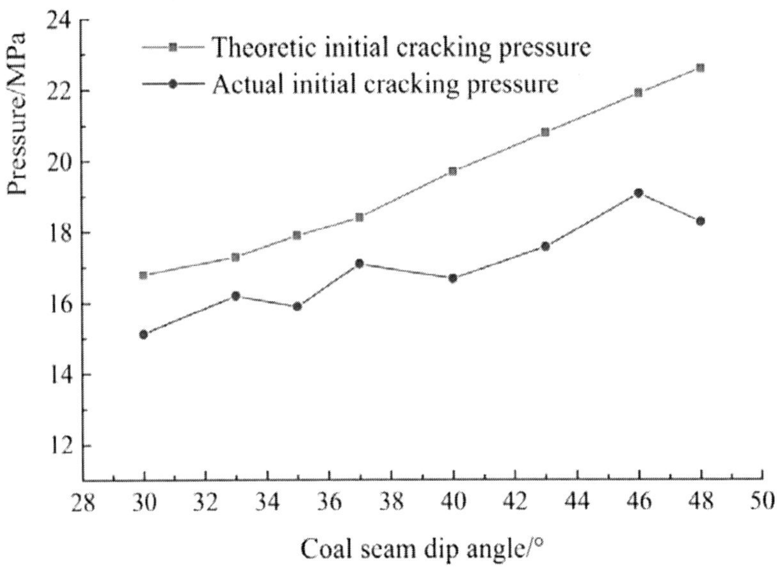


**Figure 17.** Comparison between the theoretic initial cracking pressure and the actual initial cracking pressure.

In this study, only the influences of *in-situ* stress and condition of coal seams on the initial cracking pressure were analyzed. To further reveal factors affecting the initial cracking pressure, the in-depth study will focus on the influence of rock properties and fracturing fluid on hydraulic fracturing process.

## 5. CONCLUSIONS

Using the maximum tensile stress criterion, this study analyzed the stress state around the boreholes during cross-measure hydraulic fracturing, and then developed a mathematical model for borehole wall initial cracking parameters. Finally, the influences of *in-situ* stress and coal seam dip angle on hydraulic fracturing initial pressure and locations in underground coalmines were investigated using the model. Field test data from Tonghua coalmine verified the proposed model. The results suggest that the initial cracking pressure increases with depth and coal seam dip angle, whereas it decreases with the increase in azimuth of major principle stress. This study also found that the initial cracking locations mainly occurred in the second and fourth quadrant in polar coordinate. With the increase in coal seam dip angle and azimuth of major principle stress, its direction shifted to strike of coal seam. The result of the field test is consistent with the rule that the initiation pressure increases with the coal seam dip angle, verifying the validity of the proposal mathematical model.

## ACKNOWLEDGMENTS

This study was financially supported by the National Science and Technology Major Projects of China under Grant No. 2011ZX05065, the National Natural Science Foundation of China (NSFC) under Grant No. 51374258, and Program for Changjiang Scholars and Innovative Research Team in University of China under Grant No. IRT13043.

## AUTHOR CONTRIBUTIONS

Yiyu Lu, Liang Cheng and Zhaolong Ge all contributed to developing the mathematical model, designing the experiments, and writing the paper; Binwei Xia, Qian Li and Jiufu Chen performed experiments.

## REFERENCES

1. Dai, S.F.; Ren, D.Y.; Chou, C.L.; Finkelman, R.B.; Seredind, V.V.; Zhou, Y.P. Geochemistry of trace elements in Chinese coals: A

review of abundances, genetic types, impacts on human health, and industrial utilization. *Int. J. Coal Geol.* **2012,** *94*, 3–21.

2. Shen, B.H.; Liu, J.Z.; Zhang, H. The technical measures of gas control in China coal mines. *J. China Coal Soc.* **2007**, *32*, 673–679.
3. Yuan, L. Key technique to high efficiency and safe mining in highly gassy mining area with complex geologic condition. *J. China Coal Soc.* **2006,** *31*, 174–178.
4. Lu, Y.Y.; Liu, Y.; Li, X.H.; Kang, Y. A new method of drilling long boreholes in low permeability coal by improving its permeability. *Int. J. Coal Geol.* **2010,** *84*, 94–102.
5. Karacan, C.Ö.; Diamond, W.P.; Schatzel, S.J. Numerical analysis of the influence of in-seam horizontal methane drainage boreholes on longwall face emission rates. *Int. J. Coal Geol.* **2007**, *72*, 94–102.
6. Klaus, N. Control of gas emissions in underground coal mines. *Int. J. Coal Geol.* **1998,** *35*, 57–82.
7. Sun, D.F.; Chen, J.F.; Long, J.M.; Li, W.S.; Mei, X.D. Experimental study on high pressure hydraulic fracturing technology to uncover coal in cross-cut. *Coal Sci. Technol.* **2013,** *41*, 163–165.
8. Hubbert, M.K.; Willis, D.G. Mechanics of hydraulic fracturing. *Trans. Soc. Petrol. Eng. AIME* **1957,** *210*, 153–166.
9. Schmidt, D.R.; Zoback, M.D. Poroelasticity effects in the determination of minimum horizontal principal stress in hydraulic fracturing test—A proposed breakdown equation employing a modified effective stress relation for tensile failure. *Int. J. Rock Mech. Min. Sci.* **1989,** *26*, 499–506.
10. Zhang, X.; Jeffrey, R.G.; Bunger, A.P.; Thiercelin, M. Initiation and growth of a hydraulic fracture from a circular wellbore. *Int. J. Rock Mech. Min. Sci.* **2011,** *48*, 984–995.

11. Huang, J.; Griffiths, D.V.; Wong, S.W. *In situ* stress determination from inversion of hydraulic fracture data. *Int. J. Rock Mech. Min. Sci.* **2011**, *48*, 476–481.

12. Huang, J.; Griffiths, D.V.; Wong, S.W. Initiation pressure, location and orientation of hydraulic fracture. *Int. J. Rock Mech. Min. Sci.* **2012**, *49*, 59–67.

13. Chen, M.; Jin, Y.; Zhang, G.Q. *Rock Mechanics of Petroleum Engineering*; Science Press: Beijing, China, 2008; pp. 167–173.

14. Wang, T.; Zhou, W.B.; Chen, J.H.; Xiao, X.; Li, Y.; Zhao, X.Y. Simulation of hydraulic fracturing using particle flow method and application in a coal mine. *Int. J. Coal Geol.* **2014,** *121*, 1–13.

15. Zhang, X.; Jeffrey, R.G.; Thiercelin, M. Deflection and propagation of fluid-driven fractures at frictional bedding interfaces: A numerical investigation. *J. Struct. Geol.* **2007**, *29*, 396–410.

16. Zhang, X.; Jeffrey, R.G.; Thiercelin, M. Escape of fluid-driven fractures from frictional bedding interfaces: A numerical study. *J. Struct. Geol.* **2008**, *30*, 478–490.

17. Li, X.Z.; Lin, B.Q.; Zhai, C.; Li, Q.G.; Ni, G.H. The mechanism of breaking coal and rock by pulsating pressure wave in single low permeability seam. *J. China Coal Soc.* **2013**, *38*, 918–923.

18. Lian, Z.L.; Zhang, J.; Wang, X.X.; Wu, H.A.; Xue, B. Simulation study of characteristics of hydraulic fracturing propagation. *Rock Soil Mech.* **2009**, *30*, 169–174.

19. Lin, H.X.; Du, C.Z. Experimental research on the quasi three-axis hydraulic fracturing of coal. *J. China Coal Soc.* **2011**, *36*, 1801–1805.

20. Li, C.L.; Kong, X.Y. A theoretical study on rocks break down pressure calculation equation of fracturing process. *Oil Drill. Prod. Technol.* **2000**, *22*, 54–57.

21. Hossain, M.M.; Rahman, M.K.; Rahman, S.S. Hydraulic fracture initiation and propagation: Roles of wellbore trajectory, perforation and stress regimes. *J. Petrol. Sci. Eng.* **2000**, *27*, 129–149.

22. Bradley, W.B. Failure of inclined borehole. *J. Energ. Resour. Technol.* **1979**, *101*, 233–239.

23. Aadnoy, B.S.; Rogaland, U.; Chenevert, M.E. Stability of Highly Inclined Boreholes. In Proceedings of the SPE/IADC Drilling Conference, New Orleans, LA, USA, 15–18 March 1987.

24. Tan, C.P.; Willoughby, D.R. Critical Mud Weight and Risk Contour Plots for Designing Inclined Wells. In Proceedings of the 68th Annual Technical Conference and Exhibition of the Society of Petroleum Engineers, Houston, TX, USA, 3–6 October 1993.

25. Ge, Z.L.; Mei, X.D.; Lu, Y.Y.; Xia, B.W.; Chen, J.F. Mechanical model and test study of sealed drilling for hydraulic fracturing in underground coal mines. *Rock Soil Mech.* **2014**, *35*, 1097–1103.

26. Jiang, Y.D.; Xian, X.F.; Xu, J. Research on application of Kaiser effect of acoustic emission to measuring initial stress in rock mass. *Rock Soil Mech.* **2005**, *26*, 946–950.

27. Khademian, Z.; Shahiar, K.; Gharouni, N.M. Developing an algorithm to estimate *in situ* stresses using a hybrid numerical method based on local stress measurement. *Int. J. Rock Mech. Min. Sci.* **2012**, *55*, 80–85.]

28. Jing, F.; Sheng, Q.; Zhang, Y.H.; Luo, C.W.; Liu, Y.K. Research on distribution rule of shallow crustal geostress in China

mainland. *Chin. J. Rock Mech. Eng.* **2007**, *26*, 2056–2062. (In Chinese).

29. Zhao, D.A.; Chen, Z.M.; Cai, X.L.; Li, S.Y. Analysis of distribution rule of geostress in China. *Chin. J. Rock Mech. Eng.* **2007**, *26*, 1265–1211. (In Chinese).

30. Thomas, J.L. *Fluid Jet Technology: Fundamentals and Application*; Water Jet Technology Association: St. Louis, MO, USA, 1995.

31. Zhu, B.C.; Tang, S.H.; Yan, Z.F.; Zhang, J.Z. Effects of crustal stresses and natural fractures on fracture pressure of coal reservoirs. *J. China Coal Soc.* **2009**, *34*, 1199–1202.

# CHAPTER 6

## A Standard Characterization Methodology for Respirable Coal Mine Dust Using SEM-EDX

**Rachel Sellaro [1], Emily Sarver [2] and Daniel Baxter [3]**

[1] *Department of Civil and Environmental Engineering, Virginia Tech, Holden Hall Room 108A, Virginia Tech Blacksburg, VA 24061, USA*

[2] *Department of Mining and Minerals Engineering, Virginia Tech, Holden Hall Room 108A, Virginia Tech Blacksburg, VA 24061, USA*

[3] *Environmental Analysis Associates, Inc., 5290 Soledad Road, San Diego, CA 92109, USA*

## ABSTRACT

A key consideration for responsible development of mineral and energy resources is the well-being of workers. Respirable dust in mining environments represents a serious concern for occupational health. In particular, coal miners can be exposed to a variety of dust characteristics depending on their work activities, and some exposures may pose risk for lung diseases like CWP and silicosis. As underscored by common regulatory frameworks, respirable dust exposures are generally characterized on the basis of total mass concentration, and also the silica mass fraction. However, relatively little emphasis has been placed on other dust characteristics that may be important in terms of identifying health risks. Comprehensive particle-level analysis to estimate chemistry, size, and shape distributions of particles is possible. This paper describes a standard methodology for characterization of respirable coal mine dust using scanning electron microscopy (SEM) with energy dispersive X-ray (EDX). Preliminary verification of the method is shown based several dust samples collected from an underground mine in Central Appalachia.

## KEYWORDS

Mining; Coal Dust; Respirable; Occupational Health; Particulate Composition; Dust Characterization; SEM-EDX

## 1. INTRODUCTION

A key consideration for responsible development of mineral and energy resources is the well-being of workers. Respirable dust in mining environments represents a serious concern for occupational health. Coal mine dust, in particular, has long been linked to various lung diseases like coal workers pneumoconiosis (CWP) and silicosis [1,2]. Implementation of dust regulations in the US beginning in the late 1960s has significantly decreased overall incidence of such diseases over the past several decades [2,3,4], but analysis of long-term surveillance data appears to show a recent and unexpected uptick in disease amongst some miners in particular geographic regions like Central Appalachia [3,5,6,7]. Such trends are alarming considering that most coal mines currently operate below regulatory limits on respirable dust (*i.e.*, particulates with aerodynamic diameter <10 μm), which generally pertain to total mass concentration and crystalline silica content. These trends may suggest that other exposure factors, including specific dust characteristics such as particle composition, size, and shape distributions, may be important in the occupational health context.

While MSHA's new dust rule issued in April 2014 targets further reductions in respirable dust concentrations, it is unclear if or how the lowered limits will affect health outcomes for miners in locations where causal factors for disease are not well understood. The "new dust rule"' was first proposed by the US Mine Safety and Health Administration (MSHA) on 19 October 2010 and was finally issued under the title*Lowering Miners' Exposure to Respirable Coal Mine Dust Including Continuous Personal Dust Monitors* on 23 April 2014 [8]. The rule makes a number of changes to previous regulations on dust limits and sampling in underground coal mines, and specifically will reduce the permissible respirable dust concentration from 2.0 to 1.5 mg/m$^3$. It will also require use of continuous personal dust monitors (CPDMs) by mine operators, and require that citations be issued in any instances where MSHA-collected samples for single, full shifts exceed the new 1.5 mg/m$^3$ limit.

Indeed, more comprehensive characterization of coal mine dust is necessary to fully explore these factors. Currently, a standard methodology for comprehensive, particle-level characterization of coal mine dusts does not exist. This paper describes such a methodology, which uses scanning electron microscopy equipped with energy dispersive X-ray (SEM-EDX). Although not commonly applied to respirable mine dust samples, electron microscopy with EDX has proven useful in a variety of environmental and mineral processing/metallurgical applications for fine particulate analysis. It is increasingly being used to specifically understand chemistry and morphology of airborne particulates that represent health hazards—in occupational or ambient environments. For example, methodologies for analysis of nano-sized particulates in around active welding have recently been described [9,10].

A major objective of the method development included optimization of manual analytical efforts—*i.e.*, minimizing the required SEM user time for each sample, while maximizing the range of valuable raw data types to be collected. The developed method includes particle-level analysis of composition, size and shape, from which mass and volume can also be estimated. Construction of automated spreadsheet program for computational analysis is also described here, as well as preliminary verification of the dust characterization method using three samples collected in the field.

## 2. DESCRIPTION OF DEVELOPED DUST CHARACTERIZATION METHOD

The following sections provide a detailed discussion of the particle characteristics that are included in the developed dust characterization method, as well as a description of procedures used for dust sample collection and preparation, and selection and analysis of specific particles by SEM-EDX. Additionally, computation via an automated analysis program is described for easy analysis of raw data inputs.

### 2.1. Particle Characteristics of Interest

To fully characterize particles, specific properties are of interest. Particle composition, dimensions, and shape are values which are determined with SEM-EDX, and volume and mass are calculated as a result of the analysis. These particle characteristics provide an abundance of data and information regarding respirable dust samples and aid in the comprehensive analysis of coal mine dust.

#### 2.1.1. Composition

Classification of the dust particles is based on their EDX spectra, which provides a graphical representation of the elements associated with the particle surface. The spectra are generated by detection of X-ray emissions from the particle, caused by interaction of the SEM electron beam with its surface; each element on the particle surface produces a characteristic X-ray when excited by the impinging electrons. Each peak of a spectrum, thus, represents a specific element, and relationships between peak heights can provide some indication of the elemental composition (*i.e.*, minerals can be identified by their atomic stoichiometry). For relatively small particles, such as respirable dust particulates, the electrons may penetrate deep enough into the particle (e.g., to a depth of about 1 μm) to provide relatively good information about its overall composition. However, EDX analysis on small particles is also subject to interference from the sample background (*i.e.*, if electrons penetrate completely through the particle or the electron beam is sufficiently close to the particle edge).

For the developed dust characterization method, considerable effort was aimed at establishing a set of pre-determined compositional categories into which most particles in a coal mine dust sample would be expected to fit. As a preliminary effort, lab-generated dust samples were collected using run-of-mine (ROM) coal, consisting of coal and rock (*i.e.*, primarily shale and sandstone) taken from an underground coal mine in Central Appalachia. The mine is considered "low seam" based on its average coal seam thickness of 24 inches. With an average extraction height of 40 inches, the operation is, thus, cutting about 16 inches of roof and floor rock during coal extraction.

Dust was generated under a fume hood by pulverizing a sample split from the ROM multiple times. For each dust sample collected, a pump was operated at a flow rate of 5 L/min to collect dust onto a 37 mm diameter polycarbonate (PC) filter (0.4 μm pore size), which was positioned near the top of the fume hood, just below the suction fan; this arrangement was deemed appropriate to collect relatively fine dust over short time periods (*i.e.*, 5–10 min) without the use of a cyclone or other size classifier. A cyclone was not used to collect the laboratory samples, since the primary objectives were simply determination of the ROM mineralogy (*i.e.*, such that appropriate particle composition categories could be identified), and development of standard procedures to be used during the SEM-EDX analysis. For more in-depth investigation of mineralogy, dust samples were also generated by pulverizing approximately pure rock and pure coal sub-samples hand-picked from the ROM.

An FEI Quanta 600 FEG environmental scanning electron microscope (ESEM) (FEI Company: Hillsboro, OR, USA) equipped with a Bruker Quantax 400 EDX spectroscope (Bruker Corporation: Ewing, NJ, USA) was used. In conjunction with the SEM-EDX hardware, the FEI image analysis software and Esprit EDX software provided imaging and graphical spectra results. The ESEM was operated under high vacuum at 15 kV with an ideal resolution and a working distance of approximately 12–13 mm, which was observed to be optimal for this particular scope and application. To prepare collected dust samples for SEM-EDX analysis, filters were removed with clean tweezers, and on a clean, hard surface, a 9 mm diameter trephine (*i.e.*, a cylindrical blade) and a clean razorblade were used to extract the center of the filter. The center sub-section was then attached to an SEM pin-stub mount with double-sided copper tape and sputter coated with gold/palladium (Au/Pd) to generate a thickness of about 10–20 nm (*i.e.*, 60 s sputtering time) and create the conductive surface layer needed for electron microscopy analysis.

Based on detailed analysis of the lab-generated dust samples using a number of EDX parameters, it was determined that twelve elemental peaks should be included in the developed coal mine dust characterization methodology: carbon, oxygen, sodium, magnesium, aluminum, silicon, sulfur, potassium, calcium, titanium, iron, and copper. Further, it was determined that most particles could be classified into six defined categories based on the peak height ratios: "carbonaceous", "mixed carbonaceous", "alumino-silicate", "quartz", "carbonate", and "heavy mineral". Although the ROM dust samples did not contain significant carbonate particles, carbonate particles are expected to be collected in field samples due to "rock dusting" programs in underground coal mines (*i.e.*, applying pulverized inert minerals, such as limestone or dolomite, to coal and rock surfaces underground in order to reduce explosion propagation), For the relatively few particles that could not be classified into one of these six categories, a seventh category "other" was created.

Table 1 provides examples of typical minerals associated with coal mine dust that fall into each of these categories, and defines the rules developed for compositional classification. These rules are fundamentally based on atomic abundance (*i.e.*, atomic percentage equivalencies of primary minerals in each category), which are correlated to the real-time observed peak height ratios (*i.e.*, Cps/eV) on EDX spectra of specific elements for each category. For the purpose of expedient decision making during SEM-EDX use, the observed peak heights are the main parameters used for characterization.

Each of the six defined categories has one or more dominant elements (DEs), which are associated with the mineral(s) represented that category. For a particle to be classified into a given category, the observed DE spectral peak heights must exceed the minimums shown in Table 1. It should be noted that the atomic percentage equivalents shown in Table 1 are operationally defined (*i.e.*, based significant experience of the authors and preliminary analysis of many known particle compositions), and are not representative of stoichiometry expected in the mineral(s) in each category. This is because significant interference from the filter background cannot be avoided for most particles in the respirable size range.

**Table 1.** Description of dust categories for particle classification by composition.

| Dust Category | Example Mineralogy | Parameters for Classification (Atomic % Equivalents) | Real Time Classification (Raw Peak Heights (Cps/eV)) |
|---|---|---|---|
| Carbonaceous | Coal | *Carbon* ≥ 70%<br>Oxygen ≤ 30% | *Carbon* ≥ 80<br>Oxygen ≤ 20 |
| Mixed Carbonaceous | Very thin clay minerals, or clay minerals with some carbon content | 4% > *Silicon* ≥ 2%<br>4% > *Aluminum* ≥ 2%<br>*Carbon* > 70%<br>Oxygen < 20% | 20 > *Silicon* ≥ 10<br>20 > *Aluminum* ≥ 10<br>*Carbon* ≥ 80<br>Oxygen ≤ 20 |
| Alumino-silicate | Clay minerals, feldspars | *Silicon* ≥ 4%<br>*Aluminum* ≥ 3%<br>Oxygen > 20% | *Silicon* ≥ 20<br>*Aluminum* ≥ 20<br>Oxygen > 20 |
| Quartz | Crystalline silica | *Silicon* ≥ 5%<br>Oxygen > 20% | *Silicon* ≥ 20<br>Oxygen > 20 |
| Carbonate | Calcite, dolomite | *Calcium/Magnesium* ≥ 5%<br>Oxygen > 20%<br>Carbon < 70% | *Calcium/Magnesium* ≥ 20<br>Oxygen > 20<br>Carbon < 80 |
| Heavy Mineral | Pyrite, titanium oxides | *Iron/Titanium/Aluminum* ≥ 5%<br>Oxygen > 20% | *Iron/Titanium/Aluminum* ≥ 20<br>Oxygen > 20 |
| Other | Diesel particulates, *etc.* | Does not fit any of the above | Does not fit any of the above |

Note: DE-Dominant element(s) are italicized for each defined category. For particles <1.5 μm in long dimension, DE content can be up to 50% less than the values noted in Table 1 for all defined groups with the exception of "carbonaceous". It was found that the filter media increasingly influences the spectra of smaller particles, with carbon content increasing and DE content decreasing (see below for details).

Indeed, it is well established that particles in this range often produce spectra that are influenced by electron penetration depth and/or electron scattering [11]. Electron penetration depth is generally defined as the depth at which the electron beam can penetrate the sample material. Thus, particles that are very small or thin may produce X-ray spectra that are greatly affected by filter background, and since the developed methodology for dust characterization utilizes PC filters, small particles or those with significant penetration depth are generally observed to exhibit apparently high carbon and oxygen abundances. For example, although crystalline silica particles ($SiO_2$) should exhibit a silicon and oxygen atomic percentages of roughly 47% to 53% based on stoichiometry, a respirable-sized particle on a PC background may show silicon and oxygen at 10% and 30%—with the balance being attributed carbon. Given the particle sizes in question, it is unlikely that poor liberation between materials (e.g., quartz particles ingrained in carbonaceous dust) is playing a significant role.

To further illustrate, Figure 1 shows the spectrum for a PC filter, which has atomic percentage equivalencies of approximately 85% carbon and 15% oxygen, and Figure 2 shows spectra and actual SEM images of typical "carbonaceous" and "alumino-silicate" particles. The spectra of alumino-silicates should not inherently show high abundances of carbon, but the carbon peak is observed to be very high as an artifact of PC filter interference. The phenomenon of increasing carbon content with decreasing particle size is applicable for all defined dust categories.

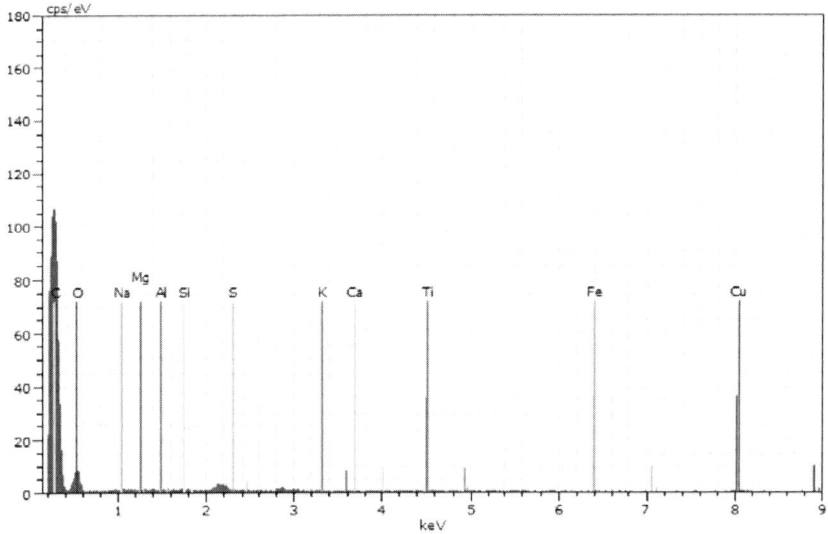


**Figure 1.** Example spectrum of the PC filter media. The red peak on the left side of the spectrum is the peak associated with carbon, and the peak to the right of carbon is the oxygen peak. The small peaks between 2 and 3 keV are the peaks from the Au/Pd sputter coating, which should be present in all spectra when Au/Pd is used to coat the samples.

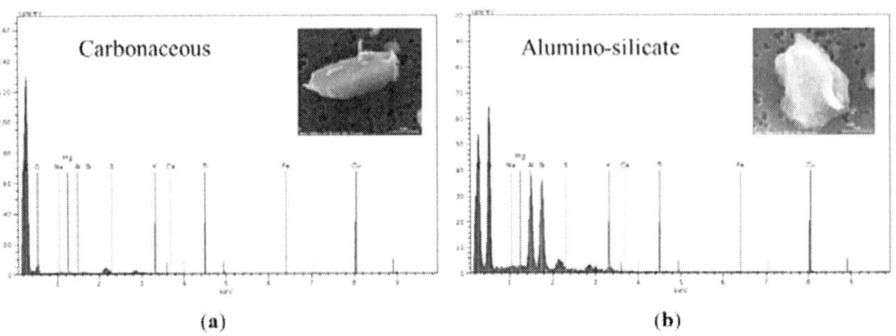


**Figure 2.** Comparison of example spectra and images for carbonaceous (**a**) and alumino-silicate (**b**) particles at 12,500× magnification. The spectrum for the carbonaceous particle (L = 9.87 μm) has a relatively large carbon peak and a much smaller oxygen peak, while the spectrum for the alumino-silicate particle (L = 11.24 μm) has relatively large oxygen and carbon peaks, and aluminum and silicon peaks of similar height.

To understand more about the particle size at which electron penetration depth may result in apparently enhanced carbon peaks, an experiment was conducted that examined quartz particles of decreasing size. To investigate, a ROM lab-generated dust sample was collected onto a PC filter. Under the SEM, the filter was scanned for quartz particles of varying sizes. Particles with a long dimension ($L$) of roughly 0.7 μm, 1 μm, 1.5 μm, 2 μm, 2.5 μm, and 3.5 μm were found, and EDX spectra were observed for each. ($L$ is simply the longest dimension visible for the particle, see e.g., [12]). Results showed that carbon peaks were higher for smaller particles; specifically, particles with $L \geq 1.5$ μm had carbon peaks <80 Cps/eV and silicon peaks >50 Cps/eV, while particles with $L <$ 1.5 μm had carbon peaks approximately 80 Cps/eV and silicon peaks approximately 20 Cps/eV. Thus, particles with $L$ much less than 1.5 μm may have exceedingly small DE peaks; and, the rules for classifying such particles into each compositional category should make allowances for their larger carbon peak due to the probability of electron penetration and/or scatter.

Understanding the carbon content of particles in the "mixed carbonaceous" category is particularly challenging. While their EDX results indicate that these particles have both alumino-silicate and carbonaceous characters, their identity and origin are not definitely known. Several possibilities exist. Most likely, particles classified as "mixed carbonaceous" are actually very thin and platy alumino-silicate particles, which are influenced ever more than other alumino-silicates by electron penetration. This prospect is supported by other recently published work by the authors [13]. Another possibility is that mixed carbonaceous particles may actually be alumino-silicates that are coated with ultrafine coal dust. Finally, it cannot be ruled out that this category could include clay mineral particles with some biogenic component, which seems possible considering the diagenesis of coal and surrounding sedimentary rock formations such as black shales.

To determine the minimum carbon content that permits classification into the mixed carbonaceous category, an experiment was conducted that looked at dust particles on a copper background media; the copper tape ensured that any electron penetration would not result in an enhanced carbon peak, but rather in copper peaks. This experiment was aimed at determining if EDX spectra from "mixed carbonaceous" particles actually exhibited high carbon peaks due to their composition, or if such peaks are simply an artifact of significant electron penetration. An ROM dust sample was collected on a PC filter, and then some of the dust particles were transferred onto copper tape and prepared for SEM analysis by the usual sputter coating routine. Particles with $L > 5$ μm whose EDX spectra exhibited relatively high aluminum and silicon peaks were specifically studied. Upon analysis of 30 such particles, only four spectra were found to have carbon peaks >80 Cps/eV. These results indicate that, in most cases, the high carbon content in "mixed carbonaceous" particles is related to interference from the PC background.

### 2.1.2. Dimensions

The long (*L*) and intermediate (*I*) dimensions of any particle analyzed can be determined directly from the SEM images using standard "line measurement" tools included in the SEM imaging software. *I* is the longest dimension perpendicular to *L*, which was defined above, in the same plane [12]. Following direct measurement of *L* and *I* (in μm), the short or third-dimension (*S*) can be estimated. Theoretically, *S* is the length dimension of a particle measured at a right angle to the plane in which *L* and *I* have been found; so *S* essentially describes particle thickness. Since different minerals have characteristic shapes, a unique ratio between *S* and *I* can usually be defined for a given mineral type. The unitless *S:I* ratio (*R*) is similar to the aspect ratio generally used in the field of sedimentology (e.g., see [14]). Alumino-silicate particles, for example, tend to be relatively flat with relatively small *R* values, whereas quartz particles tend to be thicker with higher *R* values. Thus, based on the compositional classification of each dust particle by EDX and its measured *I* value, an *S* value (in μm) can be estimated by Equation (1):

$$S = R \times I \tag{1}$$

For the particle characterization methodology developed here, the *R* values assigned to each of the six defined compositional categories of interest are as follows: 0.6 for carbonaceous, 0.5 for mixed carbonaceous, 0.4 for alumino-silicate, 0.7 for quartz, 0.7 for carbonate, and 0.7 for heavy minerals. These constants are based on those commonly used in the field of sedimentology and extensive experience of the authors in electron microscopy analysis of mineral particulates. The mixed carbonaceous category *R* value is an average of the carbonaceous and alumino-silicate values since the identity of these particles is not definitively known. Dust characterized as "other" cannot be assigned an *R* value.

### 2.1.3. Shape, Volume, and Mass

A variety of shape factors can also be computed for particles, including a measure of maximum projection sphericity ($\Psi_p$), and the cross-sectional ($d_c$) and spherical ($d_s$) diameters. The $\Psi_p$ value can be determined from the *L*, *I* and *S* dimensions using Equation (2), which was derived by Sneed and Folk (1958). $\Psi_p$ is a dimensionless quantity and values range between 0 and 1; values that approach 1 are associated with particle shapes that are increasingly spherical (*i.e.*, *L*, *I*, and *S* are very similar), whereas values that approach zero are associated with particle shapes that exhibit relatively small *S* dimensions as compared to*L* and *I* [12]. The $d_c$ and $d_s$ values (in μm) can be computed from Equations (3) and (4), respectively. The cross-sectional diameter is the only calculated value based entirely on measured properties of particle size and is only accurate if the particle is a perfect sphere. The spherical diameter is more commonly used and is considered a better approximation of the particle size in aerodynamic applications [15]. Further, the spherical volume (*V*) can also be computed (in $\mu m^3$) from Equation (5). By assigning approximate density values ($\rho$) to each compositional category, the particle masses (*m*) can additionally be estimated (in μg) using Equation (6). Based on average densities for the primary minerals expected in each of the six defined compositional categories (*i.e.*, see [16]), the following $\rho$ values (in $g/cm^3$) have been assigned: 1.4 for carbonaceous, 2.0 for mixed carbonaceous, 2.5 for alumino-silicate, 2.6 for quartz, 2.7 for carbonate, and 4.0 for heavy minerals. The mixed carbonaceous class density is an average of the carbonaceous and alumino-silicate class densities.

$$\Psi_p = \left(\frac{S^2}{L \times I}\right)^{1/3} \tag{2}$$

$$d_c = \frac{L \times I}{2} \tag{3}$$

$$d_s = \Psi_p \times L \tag{4}$$

$$V = \frac{4}{3} \times \pi \times \left(\frac{d_s}{2}\right)^3 \tag{5}$$

$$m = V \times \rho \times 10^{-6} \tag{6}$$

In addition to the shape factors noted above, particle angularity might also be considered. Angularity is an effective measure of the sharpness of the edges of a particle and, in the context of coal mine dusts, may be important in controlling interactions between respired particles and lung tissue. Angularity can be rigorously determined by measuring the observed angles of particles on SEM images; however, particularly for small particles (*i.e.*, with $L \leq 5$ μm), such analysis would require significant time. Given that a stated goal of the dust characterization method developed here was to efficiently collect data, it was, therefore, decided that a qualitative evaluation of angularity should be employed; practically, this allows for collection of some potentially valuable information without requiring excessive analytical time. This type of classification of angularity has historically been applied to particles in the micrometer size range [17,18]. To qualitatively describe angularity, particles selected for characterization should be classified as rounded (*r*), transitional (*t*), or angular (*a*) by the SEM user, see Figure 3 [19].

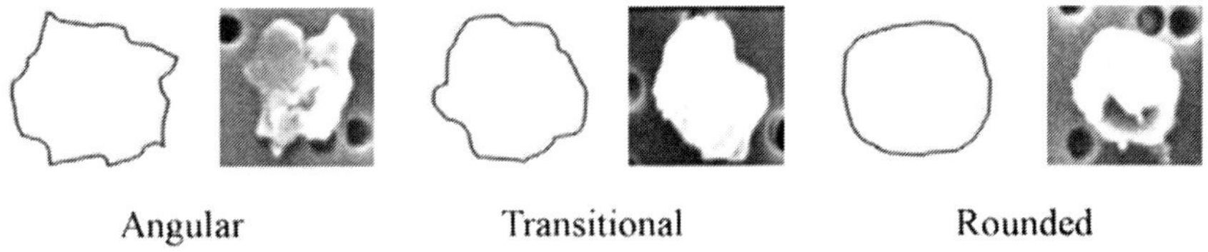


**Figure 3.** Angularity classification categories based on the qualitative analysis of the sharpness of particle edges.

## 2.2. General Procedures for Dust Characterization

In order to successfully analyze samples in a methodical manner, the collection, filter preparation, and analytical process should be sound. The following steps are outlined to provide the user with a detailed protocol to efficiently and effectively characterize respirable dust samples.

### 2.2.1. Sample Collection and Filter Preparation

For collection of respirable dust samples in the field for SEM-EDX analysis, an appropriate pump deemed permissible for use in underground coal mines must be used; at present, the MSA Escort ELF pump is almost exclusively used for such applications because it has the capability to maintain near constant flow rate under a variety of environmental conditions [20]. To ensure collection of only respirable dust particles and, thus, rejection of particles above the respirable range, the pump should be operated with a cyclone at a flow rate between about 1.7–2.2 L/min [21], such that the cyclone median cut point is 4 μm according to the NIOSH 0600 method of sampling [22]. While compliance dust samples used for determining respirable mass concentration are generally collected on pre-weighed PVC filters, samples to be analyzed by SEM-EDX should be collected on PC, because they provide a suitable substrate (*i.e.*, background media) for electron microscopy [11,23,24]. Filter cassettes should be unassembled two or three-piece types, such that the filters can be easily removed from the cassette for analysis.

In preparing the dust samples for SEM-EDX analysis, filter cassettes are carefully unassembled and the filters are removed with clean tweezers. On a clean surface, a 9 mm diameter trephine and a clean razorblade are used to extract the center of the filter. The sub-section removed for analysis represents approximately 6% of the 37 mm filter. It is recognized that particle uniformity as a function of particle size may be variable for these types of filters, which can result in larger particles depositing toward the center [25]; yet deposition is fairly radially symmetric [26]. Center filter analysis has been shown to provide reasonably precise results for field samples using two or three-piece cassettes [26]. As the main objective here is to provide relative comparisons between center filter sub-sections, some work has been completed to demonstrate that particles >0.5μm are uniformly distributed by number across the sub-section.

This filter sub-section is then attached to an SEM pin-stub mount with double-sided tape (e.g., copper, carbon), and sputter coated with gold/palladium (Au/Pd) to create the conductive surface layer needed for electron microscopy analysis. It should be noted that carbon sputter coating cannot be used since this will interfere with composition analysis by EDX of the dust particles containing carbon, but other sputter coatings (e.g., platinum, Pt) might be considered. During development of the characterization method, it was observed that a coating thickness of about 10–20 nm (*i.e.*, 60 s sputtering time) was optimal for preventing sample charging while allowing sufficient electron interaction with the dust particles to provide high-resolution SEM images and EDX spectra.

### 2.2.2. Particle Selection and Analysis by SEM-EDX

Following dust sample collection and filter preparation, SEM-EDX is used for particle characterization. Although equivalent equipment could be used, for the method outlined in this paper the same equipment and software, described above, was utilized. The developed method utilizes images obtained from a secondary electron (SE) detector for physical characterization of the dust particles (*i.e.*, to measure dimensions and qualitatively evaluate particle angularity), and EDX spectra for compositional analysis.

In order to select particles for characterization without bias, a rigorous routine was developed to navigate the prepared 9 mm diameter filter sub-sections under the SEM. The routine was developed using an iterative process, whereby over 700 particles in total from the lab-generated dust samples were interrogated for elemental composition, long and intermediate dimensions and estimated shape factors (all described in detail below). With each iteration of analysis, the routine was improved until nearly all particles encountered could be quickly classified into one of the pre-determined compositional categories described above using the EDX spectra, and raw size and shape data could be efficiently gathered for later computational analysis. It is important to note that this routine was developed based on the assumption that somewhere between 50 and 150 particles would be analyzed per dust sample, with fewer particles limiting the statistical power of results and more particles limiting practicality due to time requirements. During preliminary verification of the dust characterization method, a simple evaluation of the effect of number of particles analyzed (*i.e.*, statistical sample size) on resulting compositional distribution was conducted (see below). Ultimately, it was determined that analyzing 100 particles per sample provided enough information about the sample while maintaining reasonable analytical time requirements (*i.e.*, about 75–90 min per sample). A detailed description of the particle selection and analysis routine follows.

First, the SEM should be focused at a magnification of 10,000×, which will allow for analysis of particles within the desired size range (*i.e.*, about 0.5–8 μm); a somewhat higher magnification could be used if the particle size distribution is relatively small (*i.e.*, there are few large particles), but significantly lower magnification will prohibit adequate resolution for analysis of finer particles. With the line measurement tool, two horizontal lines are then drawn 2 μm apart and spanning the entire width of the screen, such that the space between the lines is centered on the screen (Figure 4). The SEM is then positioned such that the dust characterization will begin in the top left-hand portion of the prepared filter subsection, approximately three screen shifts from its outer edge and approximately 2.25 mm from the top (*i.e.*, one quarter of the diameter) (Figure 5). Three screen shifts from the edge of the filter prevents analysis of any particles disturbed during the filter sub-sectioning process. Additionally, the placement of the SEM stub inside the instrument determines the orientation of the "top" of the stub, based on the upper border of the screen.

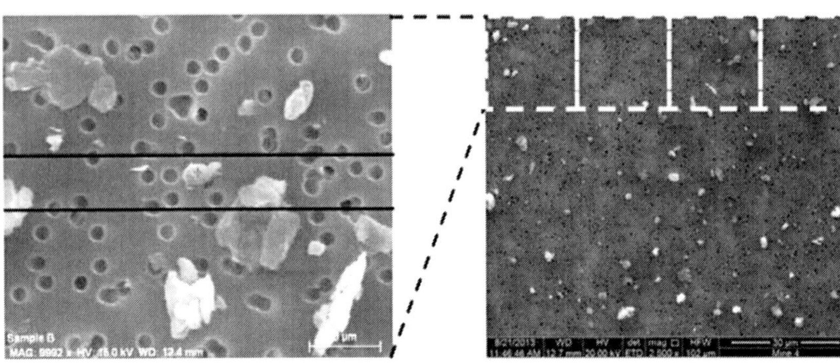

**Figure 4.** Example of particle selection and screen shifting via the joystick. The image on the left illustrates analysis of particles intersecting between the two lines in the center of the screen at 10,000× magnification. The image on the right, at 2500× magnification, shows four screens, each outlined in a white dotted line, where analysis (at 10,000× magnification) will take place consecutively.

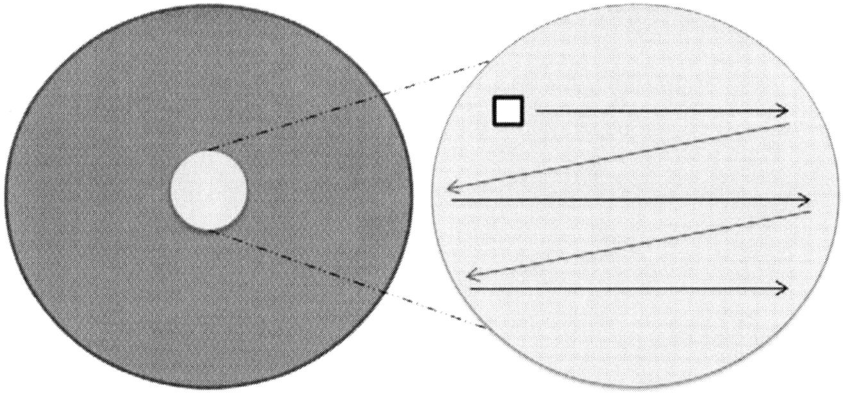

**Figure 5.** Illustration of 9 mm diameter filter sub-section and navigation routing for SEM-EDX analysis. The image on the left is the whole 37 mm diameter filter and the image on the right depicts the sub-section removed for analysis. The box in the top, left corner of the filter sub-section illustrates the first frame (*i.e.*, field of view) in which particles should be selected for characterization; the black arrows in the filter sub-section define the directions for successive screen shifts between characterization frames. When one horizontal line of analysis is complete (black arrow directions), the red arrows define shifting back to the left side of the filter to continue analysis.

Once the instrument is focused and initially positioned, selection and analysis of dust particles can begin. Moving from left to right on the screen, each particle with $L > 0.5$ μm that intersects the space between the two horizontal lines and falls completely within the field of view should be selected for analysis; if no particles in the field of view fit these criteria, the next field to the right can be examined, and so on (see below.) Particles with $L < 0.5$ μm are too small to produce quality spectra results—if analysis of smaller particles is critical, transmission electron microscopy (TEM) would be better suited for this application [27]. In an effort to analyze more of the filter area, in regards to high dust density samples, a maximum of 10 particles (*i.e.*, the first 10 that meet the above criteria moving from left to right) per field of view should be analyzed. This would allow a minimum of 10 fields of view in order to characterize 100 particles.

At the approximate center of each particle selected for analysis, the "spot" (or analogous) analysis function on the SEM software can be used with in conjunction with the EDX software to generate elemental spectra. Based on the rules outlined in Table 1, the particle can be classified into one of the seven compositional categories. Additionally, the *L* and *I* dimensions of each selected particle should be measured using the built-in line measurement tools in the SEM software. Finally, angularity should qualitatively be classified into one of the three categories described above (Figure 3). After recording raw data (*i.e.*, *L*, *I*, angularity, and composition), the user can proceed to the next particle selected for analysis. Once all eligible particles (*i.e.*, based on the criteria above) in the current field of view have been analyzed, the user should proceed to the next field of view (*i.e.*, moving to right per Figure 4 and Figure 5) for selection and analysis of more particles.

The above steps should be followed until analysis reaches the right-hand side of the filter subsection, approximately three screen shifts from its edge, or until 100 particles have been analyzed, whichever comes first. If 100 particles have not yet been analyzed, the user should navigate back to the left side of the filter subsection (see top red arrow in Figure 5), and reposition the sample such that the field of view is approximately three screen shifts from the outer edge of the filter subsection and approximately 4.5 mm from the top (*i.e.*, half of the diameter). From this position, particles should again be selected for analysis by scanning from left to right within the current field of view and adhering to the criteria outlined above; then, analysis should proceed to the next field of view. If the user again reaches the right side of the filter subsection before 100 particles are analyzed, the SEM can be repositioned back to the left—this time approximately three screen shifts from the left edge of the filter subsection and approximately 6.75

mm from the top (*i.e.*, three-fourths of the diameter). Particle selection and analysis should proceed as before.

## 2.3. Automated Analysis Program

To automate analysis of the raw data collected from SEM images and EDX spectra, a spreadsheet program was also developed using Microsoft Excel 2010 (Microsoft, Redmond, WA, USA). For each dust particle, the user inputs the compositional classification (*i.e.*, per Table 1), measured dimensions ($L$ and $I$), and qualitative angularity classification (*i.e.*, $r$, $t$, or $a$), and the program then computes the following characteristic quantities based on the assigned $R$ and $\rho$ values for each compositional category and Equations 1-6: short dimension ($S$), maximum projection sphericity ($\Psi_p$), cross-sectional diameter ($d_c$), spherical diameter ($d_s$), volume ($V$), and mass ($m$). Subsequently, distributions of composition, size (*i.e.*, $d_s$), and angularity (either by particle number or mass) can be automatically generated for each dust sample. While composition and angularity classifications are inherently categorical (*i.e.*, each particle has been placed into a specific composition or angularity category by the SEM-EDX user), particle size is continuous (*i.e.*, the computed spherical diameter is numeric quantity.) Thus, to generate distributions of quantities based on particle dimensions, a number of size categories (or classes) was defined; for this, a logarithmic base-2 scale was uses, which is a common approach used to classify particles based on work done by Wentworth (1922) [28]. Here, the automated program considers a total of nine size classes from >0.125 μm to >16 μm.

The spreadsheet program additionally includes input cells for general sample information (e.g., sample name or number, description of collection location or conditions, total filter area and filter sub-section area, total number of particles characterized, total linear length of filter analyzed), and provides basic output based on that information (e.g., percent of total filter analyzed, approximated particle density on the sub-section by mass or number). A number of graphical representations of the data results are also generated for each sample.

## 3. PRELIMINARY VERIFICATION OF DEVELOPED CHARACTERIZATION METHOD

In order to provide some preliminary verification of the characterization method developed for coal mine dust by SEM-EDX, three field samples were collected and analyzed according to the guidelines outlined above. In particular, the objectives were to: (1) verify that analysis of 50–150 particles per sample is sufficient to describe the compositional, size, and shape distributions on the filter sub-sections; and (2) verify that the six defined compositional categories using the lab-generated dust samples from ROM material, and rules for classification of particles into each category do, indeed, allow characterization of the majority of particles from real field samples (*i.e.*, do most particles fit into one of these categories, or are many particles being classified as "other"?)

It should be noted that the question of particle distribution was briefly addressed in Sellaro and Sarver, 2014. In summary, particle quantification was completed on four different areas (at 2500× magnification) of filter sub-sections from 17 field samples; this involved counting all particles with *L* dimensions >0.5 μm in each of the four areas, which were each located in a different quadrants of the filter sub-section. Particle counts were determined to be similar (*i.e.*, based on a 95% confidence interval) between each of the four areas for all but two samples. These specific filters had one quantification area with many agglomerated particles, as opposed to few, separate particles, viewed on the other three quantification areas. The agglomeration in these samples is thought to be due to humidity throughout the intake airway of the mine, where both were collected [29].

## 3.1. Materials

Three dust samples used for method verification were collected from the same underground coal mine where the ROM sample used for method development originated. An Escort ELF pump with a Dorr-Oliver cyclone was used to collect the samples onto 37 mm PC filters, and each sample was collected over a period of about 120 min. The first sample, "Roof Bolter", was collected from a location adjacent to a roof bolting machine, and thus was expected to contain relatively high proportions of alumino-silicate, and possibly quartz particles (*vs.* other compositions), due to the drilling activity of the machine into roof material. The second sample, "Belt Drive", was collected from a location just above a belt drive, where coal and rock were being transported below on a conveyor belt. The "Belt Drive" sample was anticipated to include greater proportions of carbonaceous particles, and some carbonate particles were also expected due to heavy rock dusting in the belt entries. (Rock dusting is a practice used to limit propagation of coal dust explosions, and requires walls and floors to be covered with fine inert material such as $CaCO_3$). The third sample, "Intake", was collected from a location near the working section of the mine in intake air (*i.e.*, fresh air being delivered to the mine by its ventilation system). The "Intake" sample was expected to have relatively similar proportions of carbonaceous and alumino-silicate particles, with some carbonate due to rock dusting in the area.

Estimated particle densities on the Roof Bolter, Belt Drive, and Intake filter sub-sections were 16,292 particles/mm$^2$, 12,639 particles/mm$^2$, and 1850 particles/mm$^2$, respectively. These densities were extrapolated from the average number of particles counted in four different areas on each sub-section; each area was 10,404 μm$^2$ and located in a different quadrant of the sub-section areas. Figure 6 displays SEM images for each sample.

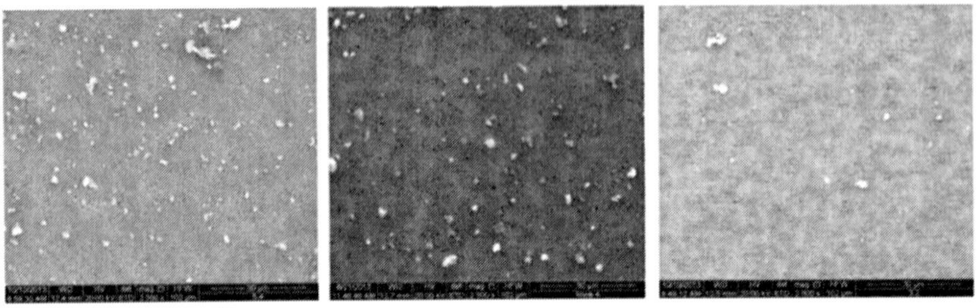

**Figure 6.** SEM images at 2500× magnification for the filter subsections from each verification sample showing relative particle densities. The far left image represents the "Roof Bolter", followed by the "Belt Drive" image, and finally the "Intake" image on the right.

## 3.2. Results and Discussion

To evaluate the effects of number of particles analyzed (*n*) on dust sample characterization results, compositional distributions by particle number and mass were compared for a range of *n* values (Table 2). For the Roof Bolter and Belt Drive samples, 200 particles in total were analyzed, and the resultant compositional distributions were compared for the first 25, 50, 100, 150, and 200 particles (*i.e.*, $n = 25, 50, 100, 150$ or 200); for the Intake sample, only 100 particles were analyzed in total, so *n* values of 25, 50, and 100 were compared. Somewhat surprisingly, when comparing compositional distribution of particles by number, all samples showed relatively similar results across all *n* values—meaning that even when *n* was increased 4- or 8-fold, little change was observed in the relative number of particles being classified into each compositional category.

When comparing compositional distribution by mass, however, only the Belt Drive sample produced similar results across all *n* values. For the other two samples, as *n* increased, the distributions changed significantly. For example, in the Roof Bolter sample, the first 100 particles analyzed showed very little carbonaceous material on a mass basis, but first 150 particles analyzed showed that over a quarter of the mass was due to carbonaceous particles. This particular discrepancy was traced to a single very large coal particle ($d_s$ > 5 μm) that was selected for analysis based on the developed dust characterization method, and it underscores a key challenge that aerosol scientists and industrial hygienists often face when studying or reporting on particulates. Indeed, due to limitations in analytical equipment and the efficiency of bulk analyses, environmental monitoring and regulation of airborne particulates is often based on mass rather than number concentrations. This analysis also suggests that in order to obtain accurate estimated mass calculations, a larger number of particles should be counted. The addition of existing automated image and X-ray analysis software could provide this capability.

**Table 2.** Distribution of particle composition by number (and mass) for the method verification samples. All values are rounded to the nearest whole number, which may result in totals being slightly different than 100%.

| **Roof Bolter** | **25 Particles** | **50 Particles** | **100 Particles** | **150 Particles** | **200 Particles** |
|---|---|---|---|---|---|
| Carbonaceous | 12% (1%) | 12% (2%) | 11% (3%) | 11% (27%) | 12% (21%) |
| Mixed Carbonaceous | 36% (31%) | 28% (25%) | 32% (19%) | 33% (26%) | 35% (23%) |
| Alumino-Silicate | 48% (68%) | 50% (64%) | 49% (67%) | 47% (39%) | 45% (41%) |
| Quartz | 4% (0%) | 8% (7%) | 6% (10%) | 6% (6%) | 7% (8%) |
| Carbonate | 0% (0%) | 0% (0%) | 0% (0%) | 0% (0%) | 0% (0%) |
| Heavy Mineral | 0% (0%) | 2% (2%) | 2% (1%) | 3% (2%) | 3% (7%) |
| Other | 0% (0%) | 0% (0%) | 0% (0%) | 0% (0%) | 0% (0%) |
| **Belt Drive** | **25 Particles** | **50 Particles** | **100 Particles** | **150 Particles** | **200 Particles** |
| Carbonaceous | 20% (27%) | 24% (40%) | 29% (37%) | 26% (36%) | 24% (27%) |
| Mixed Carbonaceous | 20% (3%) | 24% (7%) | 21% (5%) | 21% (5%) | 25% (5%) |
| Alumino-Silicate | 32% (62%) | 30% (40%) | 30% (46%) | 33% (48%) | 30% (47%) |
| Quartz | 4% (0%) | 2% (0%) | 4% (1%) | 3% (1%) | 5% (2%) |
| Carbonate | 8% (1%) | 12% (9%) | 12% (10%) | 12% (9%) | 14% (18%) |
| Heavy Mineral | 16% (7%) | 8% (4%) | 4% (1%) | 5% (1%) | 4% (1%) |
| Other | 0% (0%) | 0% (0%) | 0% (0%) | 0% (0%) | 0% (0%) |
| **Intake** | **25 Particles** | **50 Particles** | **75 Particles** | **100 Particles** | |
| Carbonaceous | 32% (56%) | 40% (50%) | 44% (44%) | 45% (48%) | |
| Mixed Carbonaceous | 20% (2%) | 16% (4%) | 17% (33%) | 16% (30%) | |
| Alumino-Silicate | 40% (22%) | 36% (25%) | 31% (12%) | 32% (14%) | |
| Quartz | 4% (1%) | 2% (1%) | 1% (0%) | 2% (0%) | |
| Carbonate | 4% (19%) | 2% (16%) | 1% (7%) | 1% (6%) | |
| Heavy Mineral | 0% (0%) | 4% (4%) | 5% (3%) | 4% (3%) | |
| Other | 0% (0%) | 0% (0%) | 0% (0%) | 0% (0%) | |

In consideration of the results presented in Table 2, and the typical times required for SEM-EDX analysis for different *n* values, it was determined that analysis of 100 total particles per sample should be both sufficient (at least for describing sample distributions by particle numbers) and practical for the developed dust characterization method. Further work may be needed, however, to determine if or how few, relatively large particles in a sample are contributing to its characteristics on a mass basis. If required, simple additional steps could be incorporated into the developed characterization method to quickly gather more data in this regard (e.g., visually scanning the filter sub-section at a relatively lower magnification to assess density of large particles, or elemental mapping at a relatively lower magnification to assess compositional differences in larger particles).

Additionally, from Table 2, it appears that the six pre-determined compositional categories, and rules outlined in Table 1 for particle classification, can account for most respirable particles in expected in dust samples from underground coal mines. Indeed, of the 500 total particles analyzed across all three samples, none required classification into the "other" category. It should of course be noted that dust composition could vary with varying coal and rock geologies, and mining and operational practices—and, thus, between mines. So, further verification of the developed method for dust characterization should certainly be conducted using samples collected from multiple mines/regions of interest.

To demonstrate the robustness of the developed dust characterization method, size, and compositional distributions (again by particle number and mass) for the three verification samples were generated by the automated spreadsheet program. Figure 7 shows the results for the sample collected adjacent to a roof bolter. With respect to composition, the sample largely consists of alumino-silicates, with significant coal and mixed carbonaceous particles too. These results are consistent with expectations based on the sampling location (*i.e.*, the bolter was drilling into the roof, but the air being moved through the mine also contains coal particles). These results additionally underscore the influence that large particles can have on mass-based data. Figure 7 indicates that 1% of the particles in this sample, which all happened to be carbonaceous, fell into the 4–8 μm size class—but these make up 19% of the total mass. Figure 8 shows the relative angularity of particles in the "Intake" sample. This data indicates that alumino-silicates and mixed carbonaceous particles tend to be primarily angular, while carbonaceous particles can be more rounded.

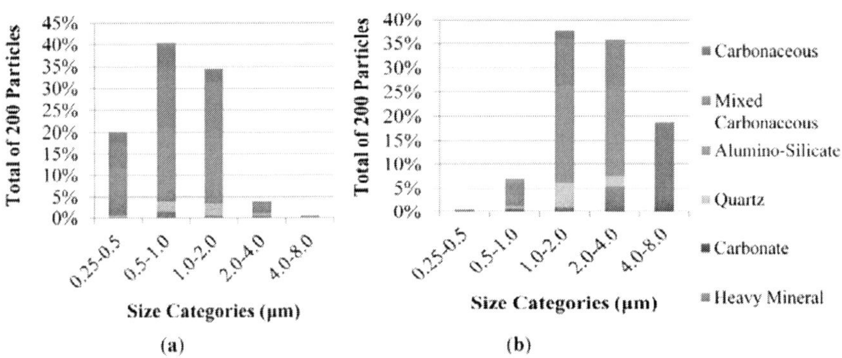


**Figure 7.** Particle size distribution by number (**a**) and by mass (**b**) for the Roof Bolter sample; the relative number of particles in each compositional category is shown within each bar.

## 4. CONCLUSIONS

SEM-EDX is a powerful tool, which can be used for particle-level analysis of dust samples. This paper describes a standard methodology developed for the purpose of achieving more comprehensive characterization of respirable dusts in underground coal mines. Due to the large amounts of data that can be generated by this method, a relatively simple spreadsheet program is recommended for automating computational analyses to compare particles within and between dust samples. The recent availability of automated particle analysis instrumentation to existing scanning electron microscopes could also provide an even more robust analysis capability by increase the number of particles analyzed by at least five to ten fold.

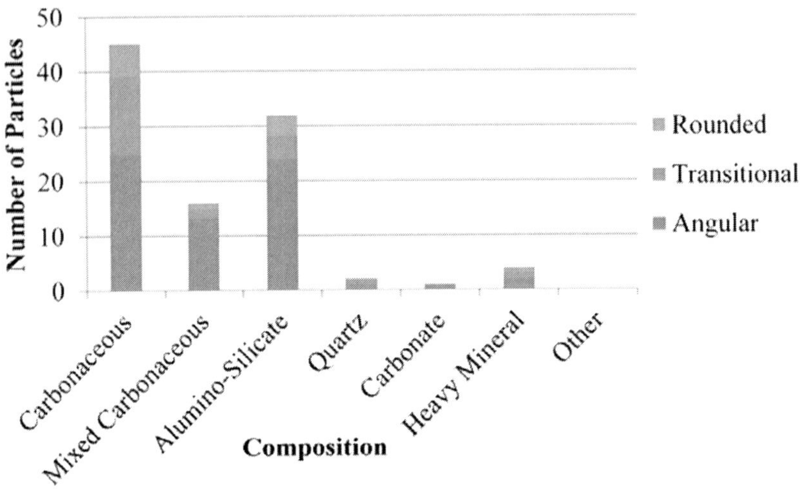


**Figure 8.** Particle compositional distribution by number for the Intake sample; the relative number of particles classified as having angular, transitional, or rounded shapes is shown within each bar.

Future work should be geared toward further understanding particle uniformity, by both number and size of particles, across the entire filter area and uniformity by particle size across the filter sub-section. In cases of non-uniformity, such as agglomerated dust, characterization of >100 particles may be necessary. The method is also user specific, and the steps outlined above are at the interpretation of the user, such as in cases of exceptionally high dust density samples and increased numbers of large dust particles. Although the method outlined in this paper was shown to classify particles properly from one specific mine, to accommodate a mine of different mineralogy, the particle dust categories should be altered prior to particle classification. The time required for this type of comprehensive analysis can be a major drawback; however, the use of a standard methodology may increase analytical efficiency, as well as consistency.

## ACKNOWLEDGMENTS

The authors would like to thank the Department of Mining and Minerals Engineering of Virginia Tech and VT NCFL for assistance with this project.

## AUTHOR CONTRIBUTIONS

Rachel Sellaro and Emily Sarver conceived the experiment and methodology. Rachel Sellaro preformed the experiment, and led the data collection, analysis and writing. Emily Sarver provided funding, guidance, and editing to the paper. Dan Baxter provided expertise, insights, and advice regarding particle analysis using SEM-EDX.

## REFERENCES

1. Attfield, M. *Current intelligence bulletin 64: coal mine dust exposures and associated health outcomes—A review of information published since 1995*; Report No. 2011–172; Department Of Health And Human Services (DHHS), National Institute for Occupational Safety and Health (NIOSH): Pittsburgh, PA, USA, 2011.

2. Laney, A.; Wolfe, A.; Petsonk, E.; Halldin, C. *Pneumoconiosis and Advanced Occupational Lung Disease among Surface Coal Miners—16 states, 2012–2011*; Report No. 61(23); MMWR: Atlanta, GA, USA, 2012; pp. 431–434.

3. Coal Workers' Health Surveillance Program (CWHSP) Data Query System. Occupational Respiratory Disease Surveillance. Available

online: http://webappa.cdc.gov/ords/cwhsp-database.html (accessed on 18 March 2014).

4. Suarthana, E.; Laney, A.; Storey, E.; Hale, J.; Attfield, M. Coal workers' pneumoconiosis in the United States: Regional differences 40 years after implementation of the 1969 Federal Coal Mine Health and Safety Act. *Occup. Environ. Med.* **2011**, *68*, 908–913.

5. Centers for Disease Control (CDC). *Advanced Cases of Coal Workers' Pneumoconiosis-Two Counties, Virginia*; Report No. 55(33). MMWR: Atlanta, GA, USA, 2006; pp. 909–913.

6. Laney, A.; Attfield, M. Examination of potential sources of bias in the US coal workers' health surveillance program. *Am. J. Public Health* **2014**, *104*, 165–170.

7. Pollock, D.; Potts, J.; Joy, G. Investigation into dust exposures and mining practices in mines in the Southern Appalachian Region. *Min. Eng.* **2010**, *62*, 44–49.

8. Lowering miner's exposure to respirable coal mine dust, including continuous personal dust monitors. Available online: http://webapps.dol.gov/federalregister/PdfDisplay.aspx?DocId=27535(accessed on 11 December 2015).

9. Gomes, J.; Albuquerque, P.; Esteves, H.; Carvalho, P.A. Notice on a methodology for characterizing emissions of ultrafine particles/nanoparticles in microenvironments. *Energy Emiss. Control Technol.***2013**, *1*, 15–27.

10. Gomes, J.; Guerreiro, C.; Lavrador, D.; Carvalho, P.A.; Miranda, R.M. TEM analysis as a tool for toxicological assessment of occupational exposure to airborne nanoparticles from welding. *Microsc. Microanal.* **2013**, *19*, 153–154.

11. Laskin, A.; Cowin, J. Automated single-particle SEM/EDX analysis of submicron particles down to 0.1 µm. *Anal. Chem.* **2001**, *73*, 1023–1029.

12. Sneed, E.; Folk, R. Pebbles in the Lower Colorado River, Texas, a study in particle morphogenesis. *J. Geol.* **1958**, *6*, 114–150.

13. Sellaro, R.; Sarver, E. Characterization of respirable dust in an underground coal mine in Central Appalachia. *Soc. Min. Metall. Explor. Trans.* **2014**, *336*, 457–466.

14. Weber, C.; Heuser, M.; Mertens, G.; Stanjek, H. Determination of clay mineral aspect ratios from conductive titrations. *Clay Miner.* **2014**, *49*, 17–26.

15. Hinds, W. *Aerosol Technology: Properties, Behavior, and Measurement of Airborne Particles*, 2nd ed.; Wiley-Interscience Publications: New York, NY, USA, 1999.

16. U.S. Geological Survey (USGS). Available online: www. usgs.gov (accessed on 8 December 2013).

17. Liebhard, M.; Levy, A. The effect of erodent particle characteristics on the erosion of metals. *Wear***1991**, *151*, 381–390.

18. Stachowiak, G. Numerical characterization of wear particles morphology and angularity of particles and surfaces. *Trobol. Int.* **1998**, *31*, 139–157.

19. Powers, M. A new roundness scale for sedimentary particles. *J. Sediment. Petrol.* **1953**, *23*, 117–119.

20. Gero, A.; Parobeck, P.; Suppers, K.; Apel, B.; Jolson, J. Effect of altitude, sampling port inlet loading, and temperature on the volumetric flow rate of the MSA Escort Elf Constant-Flow-Rate Pump. *Appl. Occup. Environ. Hyg.* **1997**, *12*, 941–946.

21. Khan, A. Characterizing the Variability in Respirable Dust Exposure Using Johnson Transformation and Re-Examining 2010 Proposed Changes to the U.S. Underground Coal Mine Dust Standard. Master's Thesis, University of Kentucky, Lexington, KY, USA, 2 May 2013.

22. National Institute for Occupational Safety and Health (NIOSH). Particulates Not Otherwise Regulated, Respirable: Method 0600. *NIOSH Manual of Analytical Methods*, 4th ed.; National Institute for Occupational Safety and Health (NIOSH): Atlanta, GA, USA, 1998; Issue 3.

23. De Bock, L.; van Malderen, H.; van Grieken, R. Individual aerosol particle composition variations in air masses crossing the North Sea. *Environ. Sci. Technol.* **1994**, *28*, 1513–1520.

24. Katrinak, K.; Anderson, J.; Buseck, P. Individual particle types in the aerosol of Phoenix, Arizona.*Environ. Sci. Technol.* **1995**, *29*, 321–329.

25. Baron, P. *Factors Affecting Aerosol Sampling*; National Institute for Occupational Safety and Health (NIOSH)/Division of Applied Research and Technology (DART): Pittsburgh, PA, USA, 2003.

26. Miller, A.; Drake, P.; Murphy, N.; Cauda, E.; LeBouf, R.; Markevicius, G. Deposition uniformity of coal dust on filters and its effects on the accuracy of FTIR analyses for silica. *Aerosol. Sci. Technol.* **2013**,*47*, 724–733.

27. Rice, S.; Chan, C.; Brown, S.; Eschbach, P.; Han, L.; Ensor, D.; Stefaniak, A.; Bonevich, J.; Vladar, A.; Walker, A.; *et al.* Particle size distributions by transmission electron microscopy: An interlaboratory comparison case study. *Metrologia* **2013**, *50*, 663–678.

28. Wentworth, C. A scale of grade and class terms for clastic sediments. *J. Geol.* **1922**, *30*, 377–392.

29. Sellaro, R.; Sarver, E. Preliminary investigation of SEM-EDX as a tool for characterization of coal mine dusts. *Min. Eng.* **2014**, *66*, 16–40.

# CHAPTER 7

# An Integrated Environment Monitoring System for Underground Coal Mines—Wireless Sensor Network Subsystem with Multi-Parameter Monitoring

**Yu Zhang [1,2], Wei Yang [1], Dongsheng Han [3] and Young-Il Kim [4]**

[1] *School of Electronic and Information Engineering, Beijing Jiaotong University, Beijing 100044, China*

[2] *Faculty of Information Technology, Beijing City University, Beijing 100083, China*

[3] *School of Electrical & Electronic Engineering, North China Electrical Power University, Baoding 071003, China*

[4] *Electronics and Telecommunication Research Institute, Daejeon 305-700, Korea*

## ABSTRACT

Environment monitoring is important for the safety of underground coal mine production, and it is also an important application of Wireless Sensor Networks (WSNs). We put forward an integrated environment monitoring system for underground coal mine, which uses the existing Cable Monitoring System (CMS) as the main body and the WSN with multi-parameter monitoring as the supplementary technique. As CMS techniques are mature, this paper mainly focuses on the WSN and the interconnection between the WSN and the CMS. In order to implement the WSN for underground coal mines, two work modes are designed: periodic inspection and interrupt service; the relevant supporting technologies, such as routing mechanism, collision avoidance, data aggregation, interconnection with the CMS, *etc.*, are proposed and analyzed. As WSN nodes are limited in energy supply, calculation and processing power, an integrated network management scheme is designed in four aspects, *i.e.*, topology management, location management, energy management and fault management. Experiments were carried out both in a laboratory and in a real underground coal mine. The test results indicate that the proposed integrated environment monitoring system for underground coal mines is feasible and all designs performed well as expected.

## KEYWORDS

multi-parameter monitoring; network management; underground coal mine; wireless sensor network

## 1. INTRODUCTION

Safety is the top priority of coal mine production and plays an important role in it. In recent years, China has gained great achievements in the field of coal mine production safety, but the situation is still severe. In China, coal mining is mainly carried out underground, where the geological conditions are complicated [1–3]. Along with coal mining at greater depths, the amount of gas emission increases, and thus the risk of coal and gas outbursts grows. Besides, there are other natural factors that can cause disasters in underground mines, such as rock bursts, coal dust explosions and water leaks. With the addition of narrow working space, poor lighting, and many hidden dangers in each coal mining procedure, heavy casualties may occur for slight carelessness. Hence, the hazardous natural conditions and the adverse working conditions make underground coal mines full of numerous risk sources and hidden dangers.

The focus of coal mine production safety is to strengthen monitoring and early warning of possible natural disasters such as gas, fires, pressure, coal dust and floods, since monitoring and controlling, predicting and warning are the precondition and guarantee for the "safety first, prevention crucial" policy, and the key to prevent accidents.

As an important guarantee for coal mine safety production, the Cable Monitoring System (CMS) technology has gone very far, and theoretical problems and the key technology issues involved have been resolved already [1]. Due to the complex working conditions in underground coal mines and the limitations of the CMS, it is difficult for the CMS itself to comprehensively and effectively monitor the environmental parameters which have important impact on the coal mine safety production, such as gas, pressure, coal dust, temperature, wind speed and carbon monoxide (CO) levels. Moreover, the area that is difficult to monitor for the CMS, is generally where the hidden dangers are most prominent, so plenty of hidden dangers are left inevitably, which is a difficult situation for the traditional CMS to overcome. Whereas, by using wireless methods, a wireless monitoring system can effectively, instantly and flexibly monitor the areas that are difficult to monitor for the CMS.

Wireless Sensor Networks (WSNs) are a new kind of *ad hoc* network, which consist of hundreds to thousands of WSN nodes that communicate with each other, and can monitor areas from small to huge [2,4]. In recent years, with the rapid development of mobile communication and high-speed electronic devices, sensors with characteristics of low power consumption, programmability, multi-parameter sensing and wireless communication ability have been put into practical use. With outstanding advantages of ease of configuration, flexibility to shrink or expand the monitoring range, strong fault-tolerance and mobility, WSNs can play an important role in monitoring and analyzing dynamic, hostile and unfamiliar environments. Therefore, based on the WSN technology, a kind of mesh WSN with multi-parameter monitoring for underground coal mines was established, acting as a supplement to the CMS. Since the WSN nodes have multiple interfaces to connect different kinds of sensors, multiple types of environmental parameters can be monitored.

Using WSNs to monitor the environment of coal mines is not a new topic [5–15]. WSN technologies for this application have been investigated before. For instance, strategies for deploying a long distance WSN in an underground coal mine have been put forwarded by Yang [7], Akyildiz has discussed the challenges of using WSNs in underground environments in his theoretical research [12], and Xiao has proposed a multi-path WSN routing protocol for mine security monitoring [14]. Most of these works are theoretical research focusing on certain technology or just focusing on the technologies of the WSN. In this paper, with the aim of practical use, an integrated monitoring system including the WSN and the CMS is constructed.

The rest of this article is organized as follows: Section 2 introduces the overall design of the integrated environment monitoring system. Section 3 introduces the supporting technologies of the WSN. Section 4 discusses network management. Section 5 presents the performance evaluation through experiments. Section 6 concludes this work.

# 2. OVERALL DESIGN OF THE INTEGRATED SYSTEM

## 2.1. System Architecture

According to the landform, a coal mine can be divided into laneways, mining areas and mined-out areas. The laneways can be further divided into main tunnels and branch tunnels [3]. In the main tunnel and most of the branch tunnels and mining areas, the landform is relatively wide, where the CMS based on an optical fiber backbone network is deployed to monitor the environment. However, the landform in some of the branch tunnels, mining areas, and all of the mined-out areas, is narrow and irregular. It is very inconvenient or even impossible to deploy CMS in these areas. In this case, WSNs are used for environment monitoring because of their flexibility [4,5]. That means the WSN is mainly deployed where is difficult to deploy the CMS. The deployment of the WSN depends on the requirements of comprehensively monitoring the coal mine, which has characteristics of regional dispersion and small relevance with individual WSN. Based on these considerations, an integrated environment monitoring system for underground coal mines is constructed, which can comprehensively monitor the coal mine environment. Figure 1 shows the architecture of the integrated environment monitoring system. As shown in Figure 1, the WSN is mainly deployed in branch tunnels, and communicates with the Ground Monitoring and Dispatching Center (GMDC) via the optical fiber backbone network.

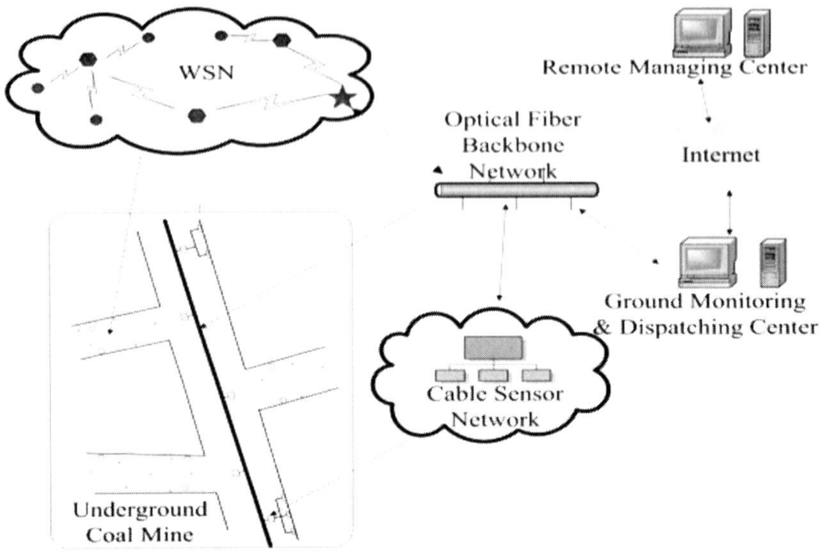


**Figure 1.** Architecture of the integrated environment monitoring system.

In this architecture, the WSN is composed of three types of nodes: sink nodes, routing nodes and sensor nodes. The sink node establishes a WSN based on the ZigBee protocol [6]. The routing node and the sensor node connect to a sink node to join a WSN. Environmental information about the underground coal mine are collected by the sensor nodes, and are sent to the GMDC through the sink nodes. According to the landform of the coal mine and the environmental monitoring requirements, different WSNs may consist of different kinds of sensors to collect different environment information. The environmental information that the CMS and the WSNs collect are transmitted to the GMDC via the optical fiber backbone network. After receiving the information, the GMDC can know the environmental conditions of the underground coal mine in real time, such as the level of gas and CO, and can rapidly respond to urgent cases. In this case, the monitoring of the environment of the underground coal mine and early warning of dangers could be achieved effectively. In addition, the connection between the GMDC and Internet make it possible for the remote manager to monitor and manage the overall safety conditions of local coal mines.

## 2.2. WSN Nodes Deployment

What is a good network topology? Based on network connectivity and coverage, by adjusting the transmitting power of nodes and selecting some nodes as backbone to deal with data processing and transmission based on certain principles, one can optimize the network topology and extend the lifetime of the whole network.

Due to the harsh environment and complex conditions of wireless communication in underground coal mines, higher requests are put forward on the stability and reliability of the WSN. According to the zonal structure of the coal mine tunnel [7–9], a kind of mesh WSN for underground coal mine is designed based on the ZigBee technology, as shown in Figure 2. In addition, the coal mine is where personnel can come to, so WSN nodes are deployed manually, which is helps optimize and manage the network.

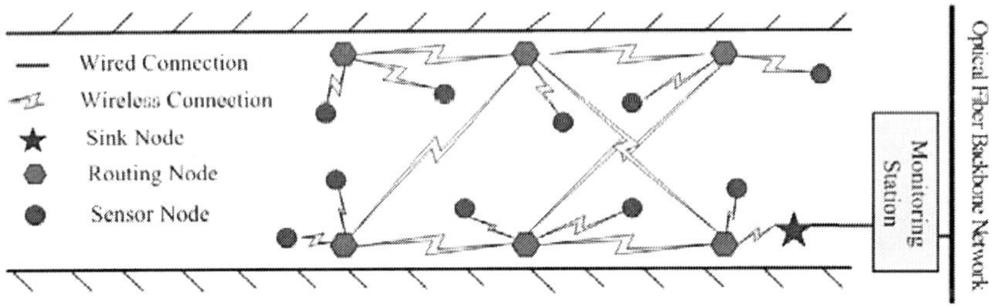


**Figure 2.** Topology structure of the WSN for underground coal mines.

As shown in Figure 2, the WSN for an underground coal mine consists of two layers: the upper backbone network and the lower perception network. The backbone network consists of a sink node and routing nodes; they can communicate with each other directly within their coverage, and mainly gather and forward data. As the gateway between the WSNs and the CMS, the sink node is deployed near the CMS cable monitoring sub-station. According to the zonal structure of the coal mine tunnels, the routing nodes are deployed along the tunnel in two symmetrical lines. In this way, every routing node can communicate with multiple routing nodes directly. If a routing node is broken, communication can still go on by using another routing node for relaying, thus the robustness of wireless communication in WSN is ensured.

The perception network consists of sensor nodes in charge of collecting environment information. They are deployed around the routing nodes and with the consideration of the environmental monitoring needs. Because the sensor node is a Reduced Function Device (RFD), it is designed only to communicate with the nearest routing node as its parent node. Besides, the transmission power of a sensor node is adjusted appropriately by using an adaptive power control mechanism. The transmission power is mainly decided by the Received Signal Strength Indication (RSSI) value between the sensor node and its parent node. By adjusting the transmission power of sensor nodes, not only is the power consumption of the sensor nodes reduced, but also the interference among neighboring nodes is reduced.

## 2.3. Working Modes

As is known, sensor nodes are powered by battery, so their working time is limited by power. In order to prolong the lifetime of sensor nodes, and at the same time ensure the environment of underground coal mine be monitored efficiently and constantly, two working modes are used in WSN, called periodic inspection mode and interrupt service mode.

### 2.3.1. Periodic Inspection Mode

Based on the deployment of the WSN nodes showed in Figure 2, the scheduling of the periodic inspection mode [11] is used, as described in Figure 3. In each cycle, the sink node starts to broadcast a parameter collection command to the routing node.

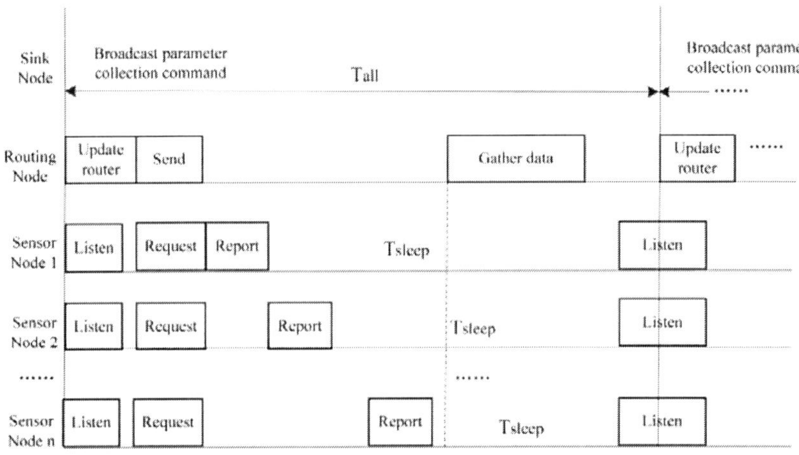


**Figure 3.** Scheduling of periodic inspection mode.

After receiving the command, the routing node first forwards it, and then determines whether it has children sensor nodes itself. If it has none, the routing node will send an active status report to the sink node after a random time period; otherwise, it will broadcast the parameter collection command to its children sensor nodes, carrying the network address of the children nodes that it has messages to transmit to, since the sensor node can only communicate with its parent node in an indirect manner. In other words, the sensor node needs to poll its parent node to determine whether it has any messages pending. When sensor node receives the command, it will first determine whether the parent node has data for it. If yes, it requests the data. After this, it collects environmental parameters according to the command, and sends them in its time slot, then goes to sleep to save power. The routing node gathers the parameters received from its children nodes, and then sends them to the sink node in a multi-hop relay way, and finally to the GMDC via the gateway. At this point, a cycle comes to an end.

### 2.3.2. Interrupt Service Mode

The natural conditions of underground coal mines are so rough that there may be a sudden incident, so it is very important for the safe production in underground coal mines to accurately monitor and provide timely alarms for sudden incidents [7]. Therefore, as the supplement of the periodic inspection mode, the interrupt service mode is designed for the WSN to warn of any abnormal situation. When the sensor node is asleep, the processor and the communication module will be shut down to save power, thus it can neither send nor receive messages, but the sensors will stay active and keep collecting environmental parameters. Once there is an overrun in a certain parameter, the sensor node will be woken up by an interrupt generated by the voltage comparator of the processor. Then, the sensor node marks the parameter as an emergency, and sends it to the GMDC immediately. When GMDC receives the emergency data, it will alarm the corresponding areas about the abnormal situation so as to predict the danger in time.

## 3. KEY SUPPORTING TECHNOLOGIES FOR THE WSN

As for the WSN, there are some technologies which are very important for its performance in underground coal mines. These technologies are called key supporting technologies, and mainly include the routing mechanism, collision avoidance, data aggregation, unified parameter gathering, network synchronization and interconnection with the CMS [12,13]. To construct an integrated monitoring system for underground coal mines, all these key supporting technologies have been discussed.

## 3.1. Routing Mechanism

As stated in Section 2.2, the WSN with multi-parameter monitoring for underground coal mines is composed of two layers. For the perception network, the sensor node can only communicate with its parent node, without routing functions. Therefore, the routing mechanism mentioned here is only applied to the backbone network.

Because the WSN nodes are limited in electric power and computing capability, in the first place the routing protocol cannot be complex. Secondly, the underground wireless communication condition is so complex that the routing protocol should be robust enough to guarantee the WSN will work stably and reliably. Thirdly, in the WSN, the sensor node mainly transmits environmental parameters, whose destination is the sink node [14–16]. Based on the above considerations, a kind of routing mechanism with minimum hops, multiple paths and based on link quality indication (LQI) is designed.

In this routing mechanism, every routing node establishes three routes, *i.e.*, parent route, minimum route and backup route, by parsing the parameter collection command which is broadcast periodically. To store the route information, every routing node maintains a routing table, containing items like *hops*, *LQI*, *flag* and*next_hop_address*, as described in Table 1. The destination is the sink node, so the destination address item is omitted.

In order to reduce the power that consumed for transmission, the hops information is carried along in the parameter collection command. The value of hops is zero when the command is first broadcast by the sink node, and adds 1 each time it passes a routing node. After receiving the command, the routing node extracts the hops and LQI information, then the link cost will be obtained based on the definition stated in the ZigBee protocol [6]. The routing node updates the routing table according to the following rules:

**Table 1.** Routing table.

| Item | Size (Byte) | Description |
|---|---|---|
| *hops* | 1 | number of routing nodes from itself to the sink node |
| *LQI* | 1 | LQI of the last communication |
| *flag* | 1 | distinguish among parent route, minimum route and backup route |
| *next_hop_address* | 2 | network address of the next hop to the sink node |

- Determine whether the command comes from its parent node. If yes, add it into the routing table as the parent route; otherwise, go to 2.
- Determine whether the minimum route is empty. If yes, add it as the minimum route; otherwise, compare it with the minimum route. If the link cost is less than that of the minimum route, replace it and replace the previous minimum route with the backup route; otherwise, go to 3.
- Compare it with the backup route. If the backup route is empty, or the link cost is less than that of the backup route, add or replace it as the backup route; otherwise, discard directly.

At this point, the routing node finishes updating the routing table in a cycle. When the routing node forwards data, it first chooses the minimum route. If it fails to send, it chooses the parent route then. If it fails to send again, it finally chooses the backup route.

## 3.2. Collision Avoidance Mechanism

According to the ZigBee protocol, WSNs that work at 2.4 GHz frequency have 16 communication channels. In order to reduce the channel interference among WSNs, orthogonal channels are assigned to the WSNs within the interference range. That means the channel of any WSN in the integrated monitoring system is orthogonal to the channels of neighboring WSNs within the interference range.

Within a WSN, all WSN nodes are working over the same channel. If they transmit simultaneously, collisions will happen, and the interference will lead to the failure of data transmission [17]. Collisions happen on both the layers of a WSN: collisions among the routing nodes in the backbone network, and collisions among the sensor nodes with the same parent routing node in the perception network.

To deal with the collisions among routing nodes in the backbone network, a kind of unslotted carrier sense multiple access with collision avoidance (CSMA/CA) algorithm is employed. According to the ZigBee protocol, the unslotted CSMA/CA algorithm means that the WSN node waits for a random time period, namely the avoidance time, before it sends data. After the avoidance time, the node starts to listen to the channel. If the channel is idle, it sends immediately; on the contrary, if the channel is busy, it waits and then tries to access to channel again.

In the WSN, a routing node normally has more than one child sensor node. When these sensor nodes send data to the routing node simultaneously, collisions will happen. For this reason, the time is divided into slots and allocated to each sensor node according to the order in which it joins the routing node. The order is computed through Equation (1):

$$n = A_n - A_{parent} - C_{skip}(d) * R_m \tag{1}$$

where, $n$ is the order in its parent node, $A_n$ is the network address of the sensor node assigned by its parent routing node, $A_{parent}$ is the address of its parent routing node, $C_{skip}(d)$ is the size of the address sub-block being distributed by each parent with depth of $d$ to its children routing node, and $Rm$ is the maximum number of routing nodes a parent may have as children.

By using this timeslot allocation mechanism, collisions among sensor nodes under the same parent routing node are effectively avoided. It also helps the routing node in gathering environmental information and managing the children sensor nodes.

## 3.3. Data Aggregate Mechanism

In the integrated monitoring system for underground coal mines, a sensor node is equipped with multiple sensors so as to collect different types of environmental parameters. To improve the transmission efficiency, the collected environment data needs to be aggregated before reporting it to the parent routing node. Therefore, a data aggregation mechanism is employed on the sensor nodes. The frame format of the aggregated data is illustrated in Table 2. Here, the *RFD* field shall be set to 0X0F, indicating that this frame is environmental information that the sensor node sends; *Limit* field shall be set to 0X01 if a parameter overruns, otherwise, be 0; *Type* field indicates the type of parameters; *Para* field carries the specific value of parameters.

**Table 2.** Frame format on sensor node.

| 1 Byte | 1 Byte | 1 Byte | Variable |
|---|---|---|---|
| *RFD(0X0F)* | *Limit* | *Type* | *Para* |

When it receives environment information from its children sensor nodes, the routing node caches it firstly. After it receives the data from all the children nodes, it aggregates it and then sends it to the sink node. The frame format of the packaged data is illustrated in Table 3. Here, the *ROUTER* field shall be set to 0X0E, indicating that this frame is environmental information that the routing node is sendsing; *Num* field indicates the number of sensor nodes in this frame; *Addr* field means the network address of each sensor node; fields of *Type and Para* have the same meaning with the corresponding items in Table 2. Because the overrun parameter is sent directly to the sink node by the sensor node, *Limit* field is not included in the frame.

**Table 3.** Frame format on routing node.

| 1 Byte | 1 Byte | 2 Byte | 1 Byte | Variable | ...... | 2 Byte | 1 Byte | Variable |
|---|---|---|---|---|---|---|---|---|
| *ROUTER (0X0E)* | *Num* | *Addr 1* | *Type 1* | *Para 1* | ...... | *Addr n* | *Type n* | *Para n* |

In the environmental information delivery process, data aggregation is adopted by the sensor node and the routing node, which decreases the amount of data, reduces data collisions, saves power, and extends the lifetime of the whole network. Meanwhile, for the overrun parameters, the sensor node directly sends the to the sink node without aggregation, realizing urgent transmission of the emergency data.

## 3.4. Unified Parameter Gathering Mechanism

As mentioned above, there are many kinds of sensors in the WSN in charge of monitoring the environment comprehensively. How to gather and recognize the different sensing data efficiently? A kind of unified parameter gathering mechanism is presented. For example, we need to collect the following eight kinds of parameters: gas, CO, coal dust, smoke, temperature, humidity, wind speed and pressure, 1-byte*Type* field in the frame as shown in Table 3 is used for expressing the type of parameters. Table 4 shows the relationship between Type-value and the corresponding parameter type, where 0x00 means that he sensor nodes are allowed to go to sleep.

As shown in Table 4, each bit of the *Type* Byte represents a kind of environmental parameter. If a kind of parameter is collected, the corresponding bit is set 1, otherwise, set 0. After receiving a parameter collection command, the sensor node starts to collect parameters according to the *Type* value in the command. If the requirement of the sensor types is more than the types of sensors with which that the sensor node is equipped, then the *Type* value should be reset according to the actual sensor types. For instance, a sensor node is equipped with two sensors: a gas sensor and a CO sensor. When it receives a command of Type value 0x07, which means collecting three gas parameters, CO and coal dust, then the *Type* value should be set 0x03. In this way, various environment parameters are gathered uniformly although different sensor nodes may be equipped with different sensors.

**Table 4.** Relationship between type-value and parameter type.

| Pressure | Wind Speed | Humidity | Temperature | Smoke | Coal Dust | CO | Gas | Type Value |
|---|---|---|---|---|---|---|---|---|
| 0 | 0 | 0 | 0 | 0 | 0 | 0 | 0 | 0x00 |
| 0 | 0 | 0 | 0 | 0 | 0 | 0 | 1 | 0x01 |
| 0 | 0 | 0 | 0 | 0 | 0 | 1 | 0 | 0x02 |
| 0 | 0 | 0 | 0 | 0 | 0 | 1 | 1 | 0x03 |
| 0 | 0 | 0 | 0 | 0 | 1 | 0 | 0 | 0x04 |
| 0 | 0 | 0 | 0 | 0 | 1 | 0 | 1 | 0x05 |
| - | - | - | - | - | - | - | - | - |
| 1 | 1 | 1 | 1 | 1 | 1 | 1 | 1 | 0xFF |

## 3.5. Network Synchronization Mechanism

The mesh WSN with multi-parameter monitoring for underground coal mines mainly works in the periodic inspection mode, but the accuracy of the crystal oscillators of the WSN nodes is generally low, which means their clocks are not accurate, so it is hard to keep WSN nodes working synchronously. In this case, a relative synchronization mechanism is proposed. In this mechanism, the sink node, which is the coordinator of the WSN, is selected as the basic benchmark node of the WSN. The other nodes select their own parent node as their benchmark node. The designed network synchronization means relative synchronization on working and sleeping, rather than absolute synchronization on a physical clock [3]. As the coordinator, the sink node is selected as the benchmark node of the whole WSN, and other WSN nodes select their own parent nodes as the benchmark node.

As designed, the sink node periodically broadcasts the parameter collection command. The command frame is also the synchronous frame in the proposed synchronous mechanism. When a WSN node receives a synchronous frame, it will check whether the synchronous frame comes from its parent. If not, it discards it directly; otherwise, it compares the sequence number of the frame with that of the last one. If they are identical, it discards the frame directly; otherwise, it records the sequence number, and regards that moment as the beginning of a new cycle.

In addition, in order to reduce the impact of clock skew, the sensor node listens to the channel both at the beginning and the end of each cycle until it receives a synchronous frame, rather than only at the beginning. In this way, the WSN nodes are capable of working synchronously without accurate clocks, ensuring the reliable delivery of environmental information.

## 3.6. Interconnection between WSN and CMS

In order to transmit the environmental information that the WSN collects to the GMDC stably, the WSN is designed to interconnect with the optical fiber backbone network via the traditional CMS, and the WSN interconnects to the CMS through the sensor node by using a Modbus protocol [18]. In addition, the sink node in the WSN connects to the CMS via a serial port. The sink node has a RS232 interface, while the CMS has a RS485 interface, so RS232/RS485 conversion equipment is used.

The Modbus protocol is a common language used by electronic controllers. It defines a message structure that controllers will recognize and use, regardless of the type of networks over which they communicate. It describes the process a controller uses to request access to another device, how it will respond to requests from the other devices, and how errors will be detected and reported.

The Modbus protocol is based on a master/slave or request/reply architecture. For our integrated environment monitoring system, the master is the cable monitoring station—it requests a message. The slave is the WSN, it takes action and responds to the master. In a Modbus network, there is one master and up to 247 slaves, each with a unique slave address from 1 to 247. For saving address, only the sink node is assigned an address for the WSN. The sink node gathers the environment information of the whole WSN, hence the information can be obtained only through the sink node.

The Modbus protocol supports two transmission modes: ASCII and Remote Terminal Unit (RTU), the RTU mode is used in the interconnection. The RTU format uses binary coding, each byte in a message contains two 4-bit hexadecimal characters, and each byte consists of a start bit, 8 data bits, a bit for even/odd parity, and a stop bit. The main advantage of this mode is that its higher character density allows better data throughput than ASCII for the same baud rate.

## 4. MANAGEMENT FOR THE WSN

In the mesh WSN with multi-parameter monitoring for underground coal mines, the WSN nodes are limited in power quantity, calculation and processing power. Moreover, they usually work without manual intervention, so some WSN nodes may fail. To ensure the stability of the WSN operation and the reliability of the data transmission, it is necessary to monitor the network health in real time, diagnose the network anomaly in a timely way, find rational solutions automatically or manually, and maintain the normal operation of the WSN [19–21]. Therefore, an integrated network management scheme is designed based on four aspects, *i.e.*, topology management, location management, energy management and fault management.

## 4.1. Topology Management

Topology discovery means acquiring the topology structure of the WSN and the physical links between WSN nodes [19]. It is the main job of topology management. A simple but effective kind of WSN topology management is proposed, whereby the mesh network topology can be obtained according to the neighbor table and the relationship between WSN nodes.

After the sink node establishes a ZigBee network, the routing node and the sensor nodes join it, and register their network address, physical address and their parent nodes' network address in the sink node. When the sink node receives this information, the tree topology can be established through the relationship between parent and child. After the WSN works steadily, the sink node knows all the routing nodes' neighbor tables by broadcasting a neighbor table request command, and the mesh topology can be established. During the operation, when nodes' status or links' status changes, the change will be reported to the sink node. According to this information, the sink node dynamically updates the topology structure, reflecting the operation situation of the WSN in real time.

## 4.2. Location Management

It is important to match the parameters that the WSN collects with their position, which benefits from analyzing environment information within specific areas and helps warn abnormal areas accurately [20]. Therefore, a kind of regional location mechanism is designed to estimate the rough location of WSN nodes in underground coal mine. For the sink node and the routing nodes, their number is not very large and their deployment is regular, so their location coordinates are configured manually. While the sensor nodes are close to the routing nodes, they are located within the parent's communication range. Then, according to the RSSI value between them, more precise locations can be obtained, reducing the complexity of network configuration.

When a WSN node lacks power or breaks down, its exact location in the underground coal mine needs to be determined, but the above location mechanism fails. Therefore, a unique identification (ID) is assigned for each WSN node, and is posted on the node. Meanwhile, their corresponding relationships are stored in the database. When the user locates a WSN node, he can rapidly determine its location according to its location coordinates, and then find the specific node based on the ID information.

## 4.3. Energy Management

The WSN nodes mainly use batteries to provide power, so their power is limited. Although some energy conservation measures are adopted, for instance, using a fixed channel allocation scheme, letting the sensor nodes work and sleep alternately and aggregating the environmental parameters, the WSN nodes still may fail for lack of power. Therefore, it is necessary to monitor the remaining power of each WSN node in real time and warn of the remaining power in a timely way [21].

After the WSN nodes join the network, they will report their remaining power to the GMDC. When the user wants to know the current remaining power, it will broadcast a power request command. After receiving the command, the WSN nodes will respond as soon as possible. In addition, based on the characteristic of processor, a function called low voltage detection is designed. When the processor detects that its voltage is lower than a given threshold (set in advance), it will generate a low voltage interruption to the GMDC. According to the warning, the user knows that a WSN node lacks power, and a technician can locate the node based on the location scheme introduced in Section 4.2 and change the batteries, ensuring the normal operation of the WSN.

## 4.4. Fault Management

Once a WSN is deployed, it works for a long time, during which a node or a link may break down, affecting the normal operation of the WSN. Although the WSNs based on the ZigBee protocol have a self-recovery function, it doesn't work in the case of hardware faults [22]. Hence, a fault management scheme is necessary to diagnose the network faults and repair them in time, so as to ensure the normal operation of the WSN.

As mentioned in Section 2.2, the mesh WSN for an underground coal mine is composed of two layers. In the backbone network, the routing nodes communicate with the GMDC via the sink node; in the perception network, the sensor nodes communicate with the sink node via the routing nodes. Based on this hierarchical structure, a hierarchy and cluster machinery is adopted to manage the WSN. The WSN is divided into two layers to manage it. In the backbone network, the sink node is the Cluster Head (CH), the routing node is the Cluster Member (CM); in the perception network, the routing node is the CH, while the sensor node is the CM. The CH is mainly responsible for managing the CMs, reporting the environmental information and the network status within its cluster to the upper network, and forwarding the commands from the upper network. By using this hierarchy and cluster machinery, management tasks are decentralized, realizing effective management for the WSN.

In the process of the WSN gathering the environmental information periodically, the information gathered by CMs must be reported via the CH to the upper network. Therefore, in each cycle, the CH knows which CMs have reported and which not. If the CH does not receive information from a CM twice, it considers that the CM may be broken. In order to determine whether the CM is really broken, the CH sends a state request command to the CM, and starts a Tcm timer. If the CH receives a reply from the CM within Tcm, it means the CM is normal; otherwise, the CH considers that the CM is really broken, and reports to the GMDC to inform the user of the failure node.

Because a sensor node can only communicate with its parent routing node, if the parent node fails, the sensor node is isolated from the sink node even if it is normal. Based on this, it is designed that if a sensor node does not receive an acknowledgement frame (ACK) from its parent node N (e.g., N = 3) times, it sets the transmitting power to maximum, and then tries to communicate with its parent. If it still fails, it considers the parent has failed, and scans the channel to rejoin the network.

According to the network status information that the GMDC receives, the failed nodes can be known and the causes can be diagnosed. Based on the location scheme introduced in Section 4.2, it is easy to locate the corresponding failure node and then remove the fault. After failure recovery, the node should be put back in its place, and its physical address should stay the same, so that the node's network address and neighbor table can be the same as before after it joins the network again, guaranteeing the stability of the network topology.

## 5. EXPERIMENT AND PERFORMANCE

In order to estimate the performances of the proposed schemes, including WSN technologies, WSN management and the interconnection between the WSN and the cable monitoring system, the corresponding tests are carried out. Firstly, experimental WSN systems are established in the laboratory environment to test the performance of the WSN technologies and WSN management schemes. Then, to test data transmission reliability of the designed WSN when it works in a real complex underground coal mine environment, tests were carried out in an actual underground mine.

Finally, tests were performed using an existing CMS to estimate the interconnection scheme between the WSN and the CMS.

## 5.1. Laboratory Experiments

A test was performed in a laboratory corridor which is about 90 m long, 2 m wide and 2.5 m high. The corridor simulates the underground coal mine laneway, while the rooms and the stairs along the corridor simulate the mining areas and the mined-out areas. As the landform of the laboratory is similar to the landform of an underground coal mine [23], tests were first carried out in the laboratory to examine the performance of the proposed WSN technologies and WSN management schemes.

When the transmission power of the WSN nodes was 0 dbm, the communication distance was about 32 m [24,25]. A WSN was established with one sink node, six routing nodes and 20 sensor nodes, and the network depth was four. The routing nodes were placed along the corridor in two symmetrical lines with an interval of 15 m, as depicted in Figure 2, and their transmission powers were all set to 0 dbm. The sensor nodes were scattered around the routing nodes randomly. There were two types: (I) one equipped with a MC112 gas sensor, a LM35 temperature sensor and a HM1 humidity sensor, which numbered 16; (II) the other equipped with a random number generator, with threshold value of 0XC0 and number of four, that means it overruns if the value is greater than 0XC0. The sink node is connected with a PC via a RS232 interface. The PC simulated the GMDC, it could display environmental parameters and information about the network state received from the WSN on a tool called serial debug terminal, and it could also publish commands to the WSN via the tool. The laboratory environment is shown in Figure 4. The working period was set to 5 min, and the working state of the network was recorded continuously for 3 h.

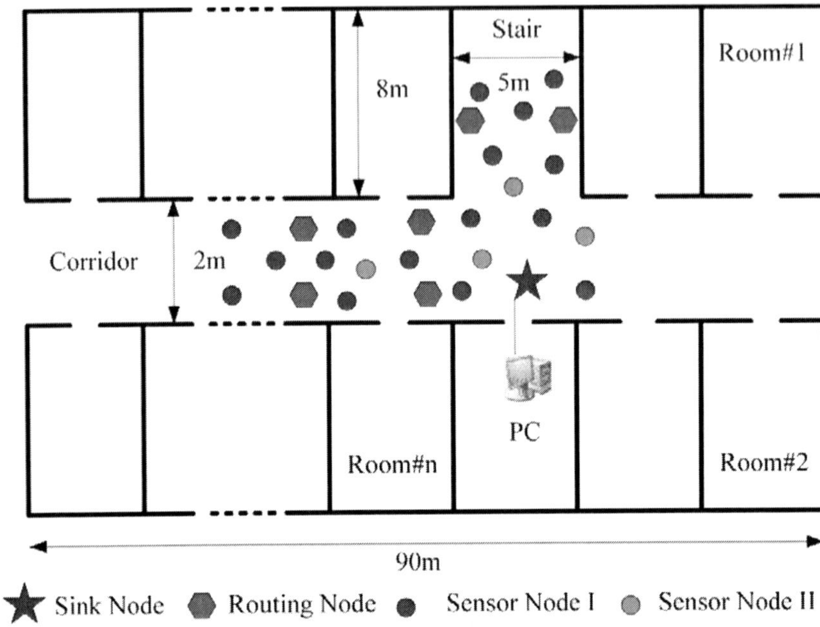


**Figure 4.** Testing in the laboratory environment.

Through a tool called Packet Sniffer running on the PC, some data frames were captured during communication. Figure 5 shows part of data frames about the routing node 0X0001 communicating with its three children sensor nodes 0X0D4E, 0X0D4F and 0X0D50, where, the sensor nodes 0X0D4E, 0X0D4F belonged to type (I), while the sensor node 0X0D50 belonged to type (II). By analysis of the data frames, information such as frame type, source address, destination address and application layer data could be parsed easily.

After the test, by calculating the number of the sensor nodes that sent each type of environmental parameter and the overrun parameter to the sink node in each period, the stability of the WSN was obtained, the statistics areshown in Figure 6.

| NWK Frame control field | NWK Dest. | NWK Src. | Broadcast | Broadcast | NWK payload |
|---|---|---|---|---|---|
| Type Version DR MF Sec | Address | Address | Radius | Seq. number | 00 D2 25 00 03 C0 D3 03 |
| DATA 0x2 1 0 0 | 0x0001 | 0x0D4E | 0x01 | 0x04 | 00 31 1A 00 2F 00 F2 00 |

| APS Frame control field | APS Dest. | APS | APS | APS Src. | APS | APS Payload |
|---|---|---|---|---|---|---|
| Type Del. mode Ind.am Sec Ack | Endpoint | Cluster ID | Profile ID | Endpoint | Counter | 00 31 1A 00 |
| DATA Unicast 0 0 0 | 0xD2 | 0x0025 | 0xC003 | 0xD3 | 3 | 2F 00 F2 00 |

(**a**)

| NWK Frame control field | NWK Dest. | NWK Src. | Broadcast | Broadcast | NWK payload |
|---|---|---|---|---|---|
| Type Version DR MF Sec | Address | Address | Radius | Seq. number | 00 D2 25 00 03 C0 D3 03 |
| DATA 0x2 1 0 0 | 0x0001 | 0x0D4E | 0x01 | 0x04 | 00 31 1E 00 29 00 26 00 |

| APS Frame control field | APS Dest. | APS | APS | APS Src. | APS | APS Payload |
|---|---|---|---|---|---|---|
| Type Del. mode Ind.am Sec Ack | Endpoint | Cluster ID | Profile ID | Endpoint | Counter | 00 31 1E 00 |
| DATA Unicast 0 0 0 | 0xD2 | 0x0025 | 0xC003 | 0xD3 | 3 | 29 00 26 00 |

(**b**)

| NWK Frame control field | NWK Dest. | NWK Src. | Broadcast | Broadcast | NWK payload |
|---|---|---|---|---|---|
| Type Version DR MF Sec | Address | Address | Radius | Seq. number | 00 D2 25 00 03 C0 |
| DATA 0x2 1 0 0 | 0x0001 | 0x0D4E | 0x01 | 0x04 | D3 03 00 02 0C 00 |

| APS Frame control field | APS Dest. | APS | APS | APS Src. | APS | APS Payload |
|---|---|---|---|---|---|---|
| Type Del. mode Ind.am Sec Ack | Endpoint | Cluster ID | Profile ID | Endpoint | Counter | 00 02 |
| DATA Unicast 0 0 0 | 0xD2 | 0x0025 | 0xC003 | 0xD3 | 3 | 0C 00 |

(**c**)

| NWK Dest. | NWK Src. | Broadcast | Broadcast | NWK payload |
|---|---|---|---|---|
| Address | Address | Radius | Seq. number | 00 CB 13 00 03 C0 D2 07 03 4E 0D 31 1A 00 2F 00 |
| 0x0001 | 0x0D4E | 0x01 | 0x04 | F2 00 4F 0D 31 1E 00 29 00 26 00 50 0D 02 0C 00 |

| APS Frame control field | APS Dest. | APS | APS | APS Src. | APS | APS Payload |
|---|---|---|---|---|---|---|
| Type Del. mode Ind.am Sec Ack | Endpoint | Cluster ID | Profile ID | Endpoint | Counter | 03 4E 0D 31 1A 00 2F 00 F2 00 4F 0D |
| DATA Unicast 0 0 0 | 0xCB | 0x0013 | 0x0003 | 0xD2 | 7 | 31 1E 00 29 00 26 00 50 0D 02 0C 00 |

(**d**)

| NWK Frame control field | NWK Dest. | NWK Src. | Broadcast | Broadcast | NWK payload |
|---|---|---|---|---|---|
| Type Version DR MF Sec | Address | Address | Radius | Seq. number | 00 CB 25 00 03 C0 |
| DATA 0x2 1 0 0 | 0x0000 | 0x0D50 | 0x0A | 0x05 | D3 04 01 02 E0 00 |

| APS Frame control field | APS Dest. | APS | APS | APS Src. | APS | APS Payload |
|---|---|---|---|---|---|---|
| Type Del. mode Ind.am Sec Ack | Endpoint | Cluster ID | Profile ID | Endpoint | Counter | 01 02 |
| DATA Unicast 0 0 0 | 0xCB | 0x0025 | 0xC003 | 0xD3 | 4 | E0 00 |

(**e**)

**Figure 5.** Data frames during communication between a routing node and its three children sensor nodes. (**a**) Sensor node 0X0D4E sending normal parameters of gas, temperature and humidity to 0X0001; (**b**) Sensor node 0X0D4F sending normal parameters of gas, temperature and humidity to 0X0001; (**c**) Sensor node 0X0D50 sending normal parameters of random number to 0X0001; (**d**) Router 0X0001 gathering parameters from (a–c);

(**e**) 0XE0 is larger than 0XC0, so 0X0D50 sends it to the sink node directly with Limit of 1As can be known from Figure 5, the WSN ran well both in periodic inspection mode and in interrupt service mode. The sensor nodes could correctly collect environmental parameters according to the command and aggregate them, and they could also report the overrun parameter to the sink node instantly. The routing nodes could correctly aggregate the parameters received from different types of sensor nodes.

As plotted in this figure, we can see that the sink node received parameters from most of sensor nodes every 5 min, meaning the periodic inspection function is stable; Meanwhile, it received some parameters from a few sensor nodes, their Limit field was 0X01 by parsing these frames, that is, they were overrun parameters, so the interrupt service function is stable. We note that the sink node did not receive parameters from all sensor nodes in some periods. Nevertheless, the number was relatively small so that we have reason to consider that the WSN works stably.

Finally, the performance on the management scheme was tested. After the establishment of the WSN, the serial debug terminal received information on network address, physical address, coordinate location, remaining power, and their parent's network address from all WSN nodes. According to the information, the network tree topology, the rough location and the remaining power of each WSN node could be obtained. Therefore, the functions of topology management and energy management are working stably. In order to test the fault management, a WSN node's electricity power was shut off manually, then the serial debug terminal received a message that a WSN node failed after several periods. By parsing the network address, the failed node was identified. Based on the failed node's coordinates and ID posted on it, it could be quickly located in the corridor. Thereby, the functions of location management and fault management are both working well.

Based on the above experiments, we conclude that the designed mesh WSN with multi-parameter monitoring for underground coal mine is easy to establish and works well. In both periodic inspection mode and interrupt service mode, the WSN can correctly collect environmental information, aggregate it, and provide early warning of overrun parameters. The proposed network management scheme is easy to implement, it can correctly reflect the network topology, diagnose the network faults in a timely way and repair them, thus ensuring the stable operation of the WSN.

## 5.2. Experiment in Real Underground Coal Mine

The field test was carried out in the Shangwan coal mine, located in Erdos (Inner Mongolia, China). The test environment is shown in Figure 7, where there is a main tunnel with a depth of about 160 m, width of 5 m and height of 4 m, and there are multiple branch tunnels along the main tunnel. Due to the underground coal mine management rules, the scale of our testing network is restricted. In this case, only two mesh WSNs were deployed: WSN_1 and WSN_2. Each WSN contains a sink node, two routing nodes and three sensor nodes. Each sensor node has three sensors: MC112, LM35 and HM1, thus is capable of collecting three types of parameters of gas, temperature and humidity, as shown in Figure 3. Figure 7b depicts the testing environment and the deployment of the WSNs.

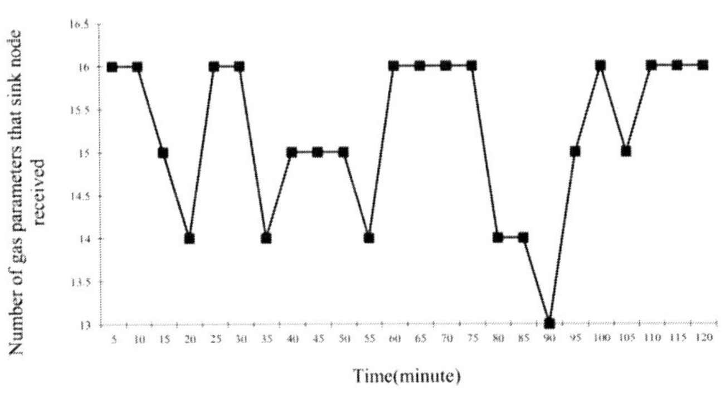


(a)

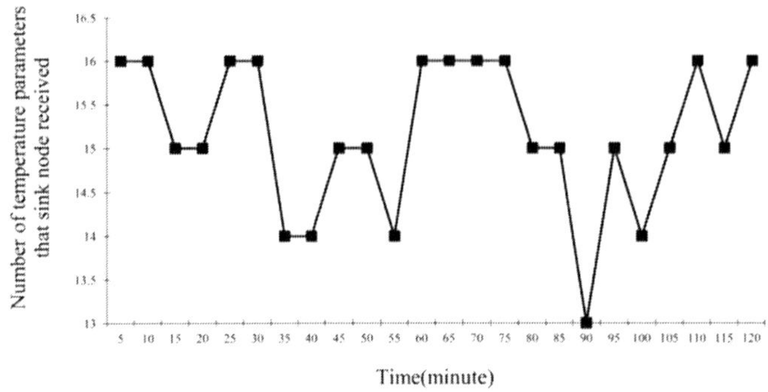


**(b)**

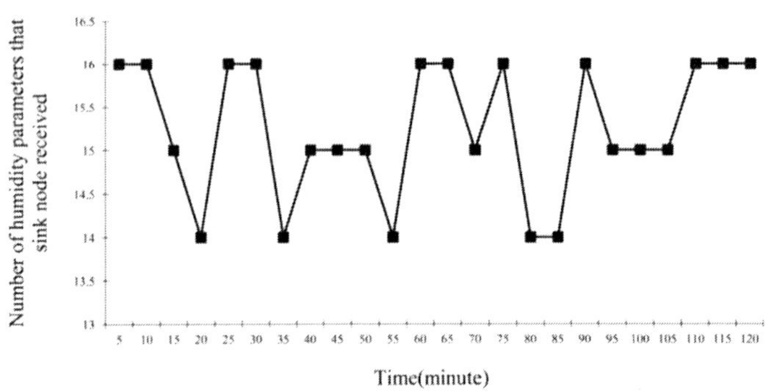


**(c)**

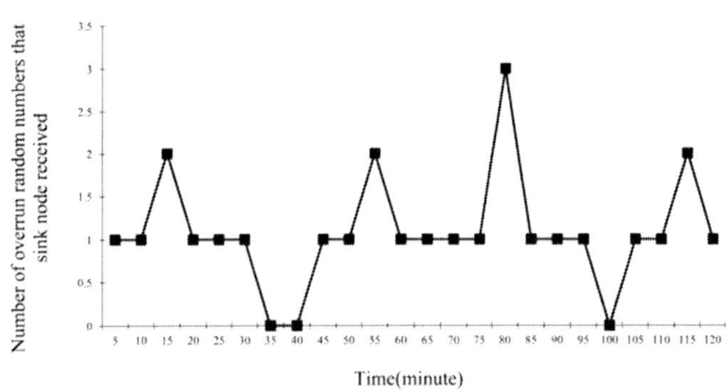


**(d)**

**Figure 6.** Statistics of each parameter in the laboratory environment. (**a**) Statistics of gas parameters; (**b**) Statistics of temperature parameters; (**c**) Statistics of humidity parameters; (**d**) Statistics of random number.

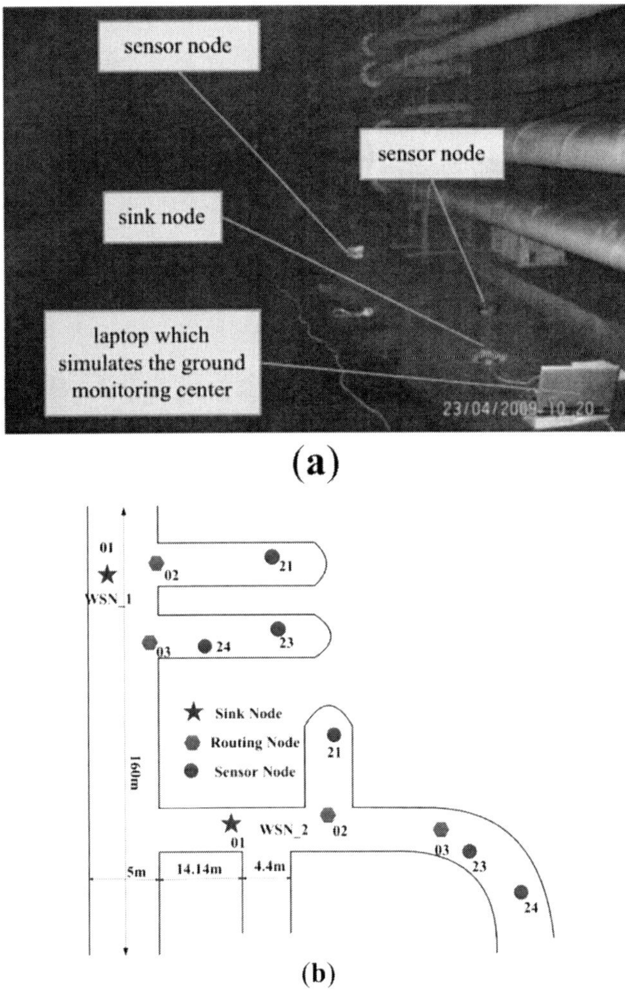


**Figure 7.** (**a**) Testing scenario in an underground coal mine; (**b**) Deployment of WSNs.

In the field test, the parameters are almost the same as the parameters of the laboratory tests, except the working period is 3 min not 5 min. To avoid influencing the normal production of the coal mine, the field test only lasted for 1 h and mainly tested the data transmission reliability.

The test results are recorded in Table 5, where, the Number of periods means the times that the sensor nodes transmitting data frames to the sink node while the Number of received data means the data frames the sink node had successfully received. As shown in Table 5, the WSNs are working reliably.

**Table 5.** Transmission results of field test. (**a**) Statistics of WSN_1; (**b**) Statistics of WSN_2.

| ID of Sensor Node | Number of Periods | Number of Received Data |
|---|---|---|
| 21 | 12 | 11 |
| 23 | 12 | 10 |
| 24 | 12 | 10 |

(a)

| ID of Sensor Node | Number of Periods | Number of Received Data |
|---|---|---|
| 21 | 18 | 17 |
| 23 | 18 | 17 |
| 24 | 18 | 16 |

(b)

## 5.3. Experiment on Interconnection between the WSN and the CMS

The interconnection testing was carried out by interconnecting the designed WSN with the KJ95N cable monitoring system which was developed by the Changzhou Institute of Automation which belongs to the China Coal Research Institute (Beijing, China). The sink node in the WSN was connected to the KJF16B monitoring station via a RS232/RS485 conversion device; and the KJF16B was connected to the KJ95N monitoring and dispatch center via a RS485 interface. KJF16B is an embedded monitoring station based on an ARM7 core processor. The environmental information of the underground coal mine is transmitted to the GMDC through the KJF16B, and the information can be displayed on its LCD. During the test, the working parameters of KJF16B were configured by the KJ95N. Three types of environmental parameters were added, including gas, temperature, and smoke. The baud rate of serial data transfer was set to 1,200 bits per second (bps).

The test data is as shown in Figure 8. The data from the WSNs is displayed on the LCD of the KJF16B monitoring station, including temperature, gas and smoke. The test results prove that the WSNs and cable monitoring system can be well connected by using the proposed interconnection scheme.

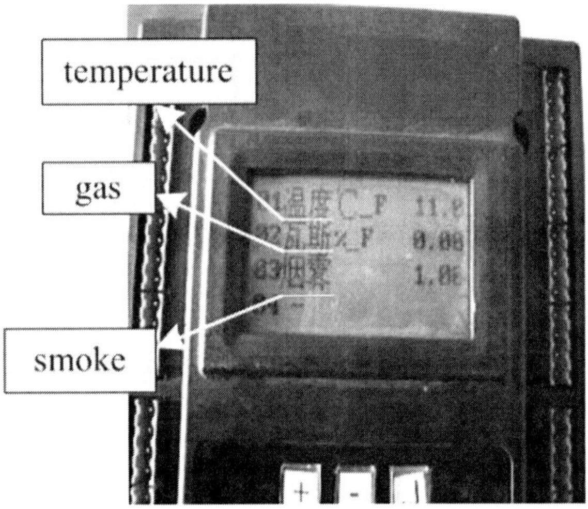


**Figure 8.** Test results displayed on the KJF16B.

## 6. CONCLUSIONS

Environment monitoring is important for the safety of coal mine production. As the traditional cable monitoring system (CMS) cannot collect all the required environmental parameters and is difficult to deploy in complex areas such as the mining areas and the mined-out areas of mines, WSN is investigated to monitor the environment in underground coal mines. An integrated environment monitoring system for underground coal mines using CMS as the backbone network and the WSN as the supplementary network is proposed. In detail, the following aspects were studied:

- The architecture of the system was designed; the deployment scheme and the working modes of WSN were put forward in detail.
- To ensure the stable and effective operation of the WSN, a set of key supporting technologies were discussed, including a routing mechanism with minimum hops and multiple paths based on LQI, a collision avoidance mechanism in combination of CSMA/CA and time division multiple access, a data aggregation strategy, a unified parameter gathering scheme and a relative synchronization scheme to make the network work synchronously. Finally, the method of interconnecting the WSN and the CMS was proposed so as to transmit the underground data of WSN to the GMDC.
- As the WSN nodes have limited energy, calculation and processing power, an integrated network management scheme was designed in four aspects, *i.e.*, topology management, location management, energy management and fault management, so as to monitor the network continuously and ensure the stable operation of the WSN.

Experiments were performed both in the laboratory and in an underground coal mine, to test the performance of the designed WSN, the proposed management scheme and the interconnection method between the WSN and CMS. The results showed that the WSN and the management scheme worked well as expected, and the interconnection between the WSN and the CMS could ensure the reliable transmission of monitoring data from the WSN to the GMDC.

The technologies used in the proposed system are all designed based on their suitability for the underground coal mine surveillance application, therefore some of them may not be so advanced in theory, but they were selected because they could make the system working smoothly as a whole. The proposed system has some good features such as universality and expansibility, and we anticipate it can be used for environment monitoring for other relevant fields.

## ACKNOWLEDGMENTS

This work was supported by the National Science & Technology Pillar Program (2013BAK06B03), Korea Research Council for Industrial Science and Technology under B551179-12-07-00, and two Natural Science Foundations of China (Grant No. 51274018 and Grant No. 61302106).

## AUTHOR CONTRIBUTIONS

The work was carried out in collaboration between all the authors. Yu Zhang, Wei Yang and Dongsheng Han defined the research theme. Yu Zhang and Wei Yang designed methods and experiments, carried out the experiments, analyzed the data, interpreted the results and wrote the paper. Young-Il Kim co-designed experiments, discussed analyses, interpretation, and presentation. All authors have contributed to, seen and approved the manuscript.

## REFERENCES

1. Ni, L.M.; Liu, Y.; Zhu, Y. China's National Research Project on Wireless Sensor Networks. *IEEE Wirel. Commun.* **2011**, *14*, 78–83.

2. Li, M.; Liu, Y. Underground Coal Mine Monitoring with Wireless Sensor Networks. *ACM Trans. Sens. Netw.* **2012**, *5*, 1–29.

3. Yang, W.; Huang, Y. Wireless Sensor Network Based Coal Mine Wireless and Integrated Security Monitoring Information System. Proceedings of the Sixth International Conference on Networking, Sainte-Luce, Martinique, 22–28 April 2007; pp. 13–15.

4. Serhan, Y.; Sabin, G.; Arslan, H.; Robin, M. Underground Mine Communications: A Survey. *IEEE Commun. Surv. Tutor.* **2009**, *11*, 125–142.

5. Shao, Q.; Zhang, Z.; Zhang, Z. Research on Safety Monitoring System of Mine Based on Ad Hoc Technology. Proceedings of the International Conference on Information Management, Innovation Management and Industrial Engineering, Xi'an, China, 23–24 November 2013; pp. 450–453.

6. ZigBee Specification. 2006. Anailable online: http:// www.zigbee.org (accessed on 2 May 2010).

7. Yang, Y.; Zhang, S.; Wang, Q.; Song, P. Long Distance Wireless Sensor Networks Applied in Coal Mine. *Procedia Earth Planet. Sci.* **2011**, *1*, 1461–1467.

8. Chen, G.; Zhu, Z.; Zhou, G.; Shen, C.; Sun, Y. Sensor Deployment Strategy for Chain-Type Wireless Underground Mine Sensor Network. *J. China Univ. Min. Technol.* **2008**, *18*, 561–566.

9. Chen, C.W.; Wang, Y.; Kostanic, I. A Chain-Type Wireless Sensor Network for Monitoring Long-Range Infrastructures. Proceedings of the International Society for Optical Engineering, Singapore, 14–16 December 2010; pp. 444–455.

10. Jang, W.; William, M.H.; Mirosław, J.S. Wireless Sensor Networks as Part of a Web-Based Building Environmental Monitoring System. *Autom. Constr.* **2012**, *17*, 729–736.

11. Zdenêk, H.; Petr, J. Energy Efficient Scheduling for Cluster-tree Wireless Sensor Networks With Time-Bounded Data Flows:

Application to IEEE 802.15.4/ZigBee. *IEEE Trans. Ind. Inform.* **2010**, *6*, 438–450.

12. Akyildiz, I.F.; Stuntebeck, E.P. Wireless Underground Sensor Networks: Research Challenges. *Ad Hoc Netw.* **2006**, *4*, 669–686.

13. Li, M.; Liu, Y.; Chen, L. Nonthreshold-Based Event Detection for 3D Environment Monitoring in Sensor Networks. *IEEE Trans. Knowl. Data Eng.* **2011**, *20*, 1699–1711.

14. Xiao, S.; Wei, X.; Wang, Y. A Multipath Routing Protocol for Wireless Sensor Network for Mine Security Monitoring. *Min. Sci. Technol.* **2010**, *20*, 148–151.

15. Chiang, S.S.; Huang, C.H.; Chang, K.C. A Minimum Hop Routing Protocol for Home Security Systems Using Wireless Sensor Networks. *IEEE Trans. Consum. Electron.* **2012**, *53*, 1483–1489.

16. Guo, Y.; Wang, Q.; Huang, H.; Tai, W.; Zhang, G. The Research and Design of Routing Protocols of Wireless Sensor Network in Coal Mine Data Acquisition. Proceedings of the 2007 International Conference on Information Acquisition, Jeju Island, Korea, 9–11 July 2007; pp. 25–28.

17. Anis, K.; Ricardo, S.; Mário, A.; Eduardo, T. Improving Quality-of-Service in Wireless Sensor Networks by Mitigating Hidden-Node Collisions. *IEEE Trans. Ind. Inf.* **2009**, *5*, 299–313.

18. Jafer, E.; Arshak, K. The Use of PE/PVDF Pressure and Temperature Sensors in Smart Wireless Sensor Network System Developed for Environmental Monitoring. *Sens. Lett.* **2010**, *6*, 477–489.

19. Yic, J.; Mukherjee, B.; Ghosal, D. Wireless Sensor Network Survey. *Comput. Netw.* **2008**, *52*, 2292–2330.

20. Ruiz, L.B.; Nogueira, J.M.; Loureiro, A.F. MANNA: A Management Architecture for Wireless Sensor Networks. *IEEE Commun. Mag.* **2003**, *41*, 116–125.

21. Zhang, B.; Li, G. Survey of Network Management Protocols in Wireless Sensor Network. Proceedings of the International Conference E-Business and Information System Security, Wuhan, China, 23–24 May 2009; pp. 1–5.

22. Asim, M.; Mokhtar, H.; Merabti, M. A Fault Management Architecture For Wireless Sensor Network. Proceedings of the Wireless Communications and Mobile Computing Conference, Cagliari, Sardinia, Italy, 1–5 July 2013; pp. 779–785.

23. Niu, X.; Xi, H.; Zhao, Z.; Zhang, Y.; Huang, C.; Cui, L. The Design and Evaluation of a Wireless Sensor Network for Mine Safety Monitoring. Proceedings of the IEEE Global Telecommunications Conference, Anaheim, CA, USA, 3–7 December 2012; pp. 1291–1295.

24. Chehri, A.; Hussein, M.; Fortier, P.; Aniss, H. Experimental Testing of IEEE801.15.4/ZigBee Sensor Networks in Confined Area. Proceedings of the 8th Annual Conference on Communication Networks and Services Research, Montreal, QC, Canada, 11–14 May 2010; pp. 244–247.

25. Ruiz-Garcia, L.; Barreiro, P.; Robla, J.I. Performance of ZigBee-Based Wireless Sensor Nodes for Real-time Monitoring of Fruit Logistics. *J. Food Eng.* **2011**, *87*, 405–415.

# CHAPTER 8

## Estimating Limits for the Geothermal Energy Potential of Abandoned Underground Coal Mines: A Simple Methodology

**Rafael Rodríguez Díez * and María B. Díaz-Aguado**

*Oviedo School of Mines, University of Oviedo, Independencia 13, Oviedo 33004, Spain*

## ABSTRACT

Flooded mine workings have good potential as low-enthalpy geothermal resources, which could be used for heating and cooling purposes, thus making use of the mines long after mining activity itself ceases. It would be useful to estimate the scale of the geothermal potential represented by abandoned and flooded underground mines in Europe. From a few practical considerations, a procedure has been developed for assessing the geothermal energy potential of abandoned underground coal mines, as well as for quantifying the reduction in $CO_2$ emissions associated with using the mines instead of conventional heating/cooling technologies. On this basis the authors have been able to estimate that the geothermal energy available from underground coal mines in Europe is on the order of several thousand megawatts thermal. Although this is a gross value, it can be considered a minimum, which in itself vindicates all efforts to investigate harnessing it.

## KEYWORDS

geothermal energy; mine water; coal mines; abandoned mines

## 1. INTRODUCTION

Large mining areas in Europe are currently being affected by closure processes, which are mainly due to the progress in mining works and changes in mining activities. Mine closure creates negative social, economic, urban and environmental effects on the affected areas. Although mines present high potential for geothermal utilization of low-temperature water, which could be used for heating and cooling purposes, only a few cases have been reported in Europe, Canada, USA or China where this potential geothermal energy from underground mines has been actually detected and used.

In Europe, some cases from The Netherlands, Germany, Poland, United Kingdom, Norway and Spain have been reported by several authors [1,2,3]. Power obtained from mine water can only be a few kilowatts thermal (kWt) from small installations, like Freiberg (Germany) or Shettleston and Lumphinnans (Scotland, UK); but there are also large installations which extract several megawatts thermal (MWt) from mine waters like those in Heerlen (Netherland), Mszczonow (Poland), and Mieres (Spain). Therefore, a vast geothermal potential is not being exploited; nevertheless, there is always a problem before the start of any project using the geothermal power of a mine. The procedure begins when there is an abandoned mine, as in the two cases shown in Figure 1 (one is an old gallery, and the other is an old shaft). These photos illustrate the reality of the initial stage of these projects: just evidence of the existence of two underground mines which were in operation in the past.

(A)

(B)

**Figure 1.** (**A**) Abandoned underground mine, Mariana mine (Asturias); (**B**) Abandoned underground mine, Olloniego colliery (Asturias).

Without doubt, there are strong reasons related to sustainability and ecology which make this kind of projects worthy of research; nevertheless, economic viability is always a strong point helping support the project. Thus, the first question that should be solved is: could the use of geothermal power from these mines be profitable? When revising the specialized literature, not many cases have been described. Relevant research has been carried out mainly in Canada [4,5,6,7,8,9,10] (the Springhill case is of special interest because it has been operating since 1988) and Europe [11,12,13,14,15,16,17,18]. Another few more cases of geothermal use of mines have also been reported concerning mines in other countries such as the USA [19,20] and China [21,22,23]. An interested reader can find more complete information about these cases in [1,2,3] and [16]. In many of these works, hydrogeological models are described. In any cases, a great deal of information is needed to conduct these studies successfully. When the mine is in operation, to find this information is easy. On the contrary, it is difficult to obtain such information when the mine was closed many years ago and the company has disappeared. It is often necessary to study old public administration documents or even visit historic libraries or registries, making this an archaeological and/or industrial patrimony task rather than technical research. Such an amount of work would be only justified in those cases where a significant geothermal energy extraction is expected. The same question can be asked at another level. For example, in countries having a long mining tradition, is it reasonable to carry out actions related to geothermal use from abandoned mines which imply a large investment? How much geothermal energy is expected to be extracted from abandoned mines? In this sense, the main objective of this research work is to develop a method which allows estimating lower and upper limits easily for the geothermal potential of an abandoned underground coal mine. This is important in order to evaluate the possibility of an actual geothermal use of an abandoned mine or of a future use for a mine near closure. The method would help to decide under which conditions it becomes interesting to start a project for using this energy from mines.

In this work, the "geothermal potential" of a mine is the total amount of geothermal energy (or geothermal power) which can be obtained from this mine. It is easy to understand that one of the factors influencing geothermal potential is the volume and characteristics of the voids created by mining activity (*i.e.*, these voids can be stable in the long run and remain open, or they can cave in immediately and be filled with rock debris). This value is directly related to coal output. Our goal is to relate the present geothermal potential of an underground coal mine with the total saleable coal production yielded by the mine through its operation history. The main advantage of this proposal is that the coal output of a mine is a well-known parameter, which is always easy to find because it has been recorded over the years by public administration. On the other hand, wide coal mining experience helps us to establish an easy-to-use method.

Thus, a simple formula is proposed:

[illegible] (1)

where $W_t$ is the value for geothermal power of the mine in MW thermal (MWt); $P_T$ is the total saleable coal production in millions of tonnes (Mt); and k is the factor of proportionality which has to be estimated empirically.

Before starting a project for the use of geothermal power of an abandoned mine, it would be very interesting to make a "reasonable estimation" of the minimum and maximum thermal energy which can be recovered from mine water. If the minimum quantity is sufficiently high, a project for its use could be proposed. Nevertheless, under some unfavorable conditions (as for example a mine located far away from inhabited areas) the project could be rejected if it were not economically feasible, even assuming maximum heat recovery.

A "reasonable" maximum and minimum value for the ratio $W_t/P_T$ will be determined in the following. Regarding our mines, the authors have determined that $k_{min} \approx 0.25$ and $k_{max} \approx 1.0$. It is not possible to make accurate predictions using this methodology; nevertheless, if technicians were able to predict a minimum value (or lower limit) for the geothermal potential of a mine and this value is high enough, the development of a project can be justified. On the other hand, if a maximum value of geothermal power is calculated and it is not sufficiently high to support a project, it is clear that it is better not to spend resources in this project.

In order to define this empirical method, coal output and two other parameters of the mine also have to be known: maximum water pumping and average quantity of air flow. Two different mines (different coalfields, history, exploitation methods, hydrogeological conditions...) are described here below. An analysis of these mines helps to understand the method and the value of the characteristic parameters, $k_{min}$and $k_{max}$.

It has to be pointed out that the method is only proposed for underground coal mines. As is well known, due to its sedimentary origin, coal usually appears in Nature as coal seams. This fact makes underground coal mines have a more topologically arranged structure than base-metals mines. Consequently, it is easy to find relationships between different parameters which allow the final relationship between geothermal power and total coal ouput to be found. For example, in a given coal field, the total length of galleries is approximately proportional to coal production. In base-metals mines it is more difficult to establish these kinds of relationships due to variable metal concentration in the rockmass. This does not mean that in this kind of mines it could not occur; in fact, a similar formula could probably be established for base-metals mines, but since no data is available a similar empirical law cannot be defined.

## 2. BRIEF DESCRIPTION OF THE PROPOSED METHOD

As previously mentioned, the method is reduced to a simple formula (Equation (1)). This formula allows estimation of the geothermal power of the mine $W_t$ (in MW thermal or MWt), from a well-known parameter which is the total saleable coal output $P_T$ (in millions of tonnes or Mt) produced by this mine during operation period. In effect, $P_T$ is easy to find, since it has always been recorded over the years by different administrations. In order to define the method, the parameter k has to be defined or estimated, which implies knowing both the geothermal power and coal output of a mine at a given moment.

In any active mine, there is always a constant flow of two fluids which interchange heat with the rockmass: water and ventilation air. An estimation of the geothermal power which could be supplied by the mine can be deduced from the total heat extracted by these two fluids from the mine.

Assuming that a certain quantity of water flow $Q_w$ ($m^3/s$) is pumped from the mine and assuming that the temperature of the water has increased in $\Delta T$ (°C) in its flow through the rockmass, the heating power $W_w$(Watts) which heats the water is:

$$W_w = Q_w \times d_w \times s_w \times \Delta T_w \qquad (2)$$

$d_w$ and $s_w$ are respectively the density ($kg/m^3$) and the specific heat (J/kg·°C) of water.

In the same way, if the air flow rate $Q_a$ ($m^3/s$) is extracted from the mine by the main exhaust fans and its temperature has increased by $\Delta T_a$ (°C), the heating power $W_a$ which heats the air is:

$$W_a = Q_a \times d_a \times s_a \times \Delta T_a \qquad (3)$$

where $d_a$ and $s_a$ are respectively the density and the specific heat of air.

Nevertheless, the increase of water and air temperature can be produced by other heat sources in the mine which are not related to the heating capacity of the ground. In underground coal mines, the most important artificial heating source is the electrical equipment which also contributes to increasing the water and air temperature.

Assuming a total electrical power E (MW) installed in the mine and an electrical performance of r (%), the total power transferred to the air/water would be:

$$W_e = (100 - r) \times E \tag{4}$$

Under these conditions, the total thermal power effectively released from the rockmass or transferred from the rockmass to water and air is:

$$W_{tmin} = W_w + W_a - W_e \tag{5}$$

This is a real value, since it has been directly obtained from experience and, without doubt, the available geothermal power of this mine is at less equal to it.

In Asturias, the average temperature of water pumped from mines is about 18 °C [24,25] whereas the average temperature of water at the surface is around 12 °C, and consequently, $\Delta T_w \approx 6$ °C. On the other hand, the water density and specific heat are $d_w = 1000$ kg/m$^3$ and $s_w = 4186$ J/kg·°C respectively. The heating power which heats the water, which is a part of geothermal power of the mine, can be estimated by:

$$W_w = 25.1 \times Q_w \tag{6}$$

where $W_w$ is in megawatts thermal (MWt) when $Q_w$ is in cubic meters per second (m$^3$/s).

In winter, the average temperature of mine air is also about 18 °C and the temperature of the air at the surface is about 7 °C. On the other hand, taking into account that the air humidity within the mine is almost constant and close to 90%, its density and specific heat are 1.18 kg/m$^3$ and 1020 J/kg·°C respectively; the thermal power necessary to heat the air is thus:

$$W_a = 0.013 \times Q_a \tag{7}$$

$W_a$ is in megawatts thermal (MWt) when $Q_a$ is in cubic meters per second ($m^3/s$).

On the other hand, assuming an electrical performance of 90%, the total power is:

$$W_e = 0.10 \times E \tag{8}$$

Under these conditions, the total thermal power released from the rockmass or transferred from the rockmass to water and air is:

$$W_t = W_w + W_a - W_e = 25.1 \times Q_w + 0.013 \times Q_a - 0.10 \times E \tag{9}$$

Taking into account the previous explanation, the ratio can be easily calculated as:

$$k = W_t/P_T \tag{10}$$

The analysis of two mines where this ratio reaches low and high values respectively allows us to estimate a minimum and maximum value for k.

The method is useful to perform geothermal resource estimates for given mining regions where coal extraction data are available; nevertheless, it is important to point out that it should not be used to design a geothermal system at a mine site.

The most important factor when using geothermal energy from flooded mines is that there must be a customer for the energy nearby. However, villages or even towns and cities have typically grown due to a mine having started its mining activity nearby. Consequently, a lot of mines in Europe are near populated areas and the geothermal energy can be used directly in district heating or similar systems.

# 3. EMPIRICAL ESTIMATION OF LIMIT VALUES FOR PARAMETER K

## 3.1. Case History 1: La Camocha Colliery

This mine has exploited an independent coal field in the past. Coal seams were mainly very steep (70° dips) and of low-medium thickness (1.5 m in average). The mining method used initially was the traditional inverted steps method (with backfilling) when mining was manual by means of pick hammers. More recently, sublevel caving with explosives has been used successfully.

This is an example of estimating the geothermal power of abandoned coal mine from historic coal output data, air flow rate and quantity of water pumped. As the quantity of water pumped out of this mine can be considered rather low, data will be used for the estimation of a minimum value $k_{min}$.

In this case, coal output data are known from the first year of the mine until the last year of exploitation. Annual coal production of the mine during its history is represented in Figure 2A.

The exploitation of the mine started during the 1930's. During the Second World War in Europe, the price of coal increased and this caused an increase in coal production. The maximum was reached in 1960 and then, output decreased quickly mainly due to emigration of miners to other coal fields in Europe which offered better working conditions (in this case, to Germany and Belgium). This tendency continued until 1970. However, the energy crisis in the 70s made the production of coal interesting and coal output increased again. Such an increase continued until about 1995. Then, changes in the world market and in European politics caused coal output to decrease drastically until mining ceased, in 2007. Until then, total accumulated production was about 16 million tonnes of saleable coal (Figure 2B).

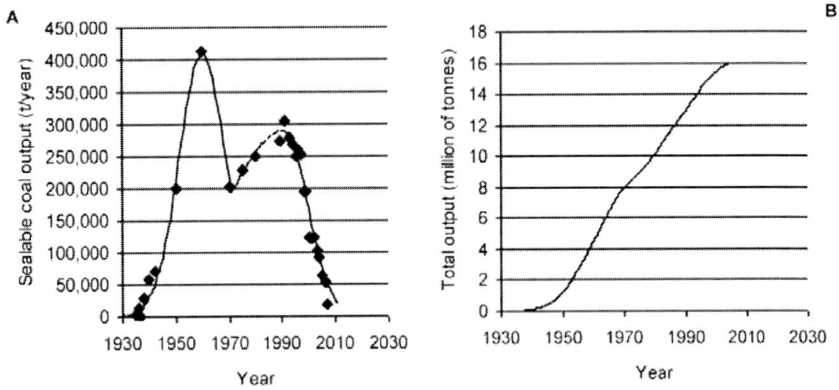


**Figure 2.** Coal production at La Camocha Colliery (Asturias, Spain). (**A**) Coal output, in t/year; (**B**) Cumulative total output from approximately 1932 to mine closure.

The use of this method implies knowing not only coal output but also another two parameters of the mine: water pumping and quantity of air flow. It is not always possible to obtain the necessary information; nevertheless, in this case, a record for the water pumping over several years has been obtained (Figure 3A).

This is a characteristic curve which decreases over time. There is a initial period of transient regime when the water originally contained in rockmass flows towards the mine. Afterwards, a stationary regime is reached (the water inflow into the mine is equal to the water inflow within the rockmass) and the quantity of water flow remains more constant during the years. The greatest water flow rate pumped was 200 $m^3/h$.

As can be seen from Figure 3B, water pumping is related to coal output because underground voids created by mining works become channels for water flow. On the other hand, ground movements caused by caving and land subsidence generate an increasing rock mass permeability. But, the relationship between water and coal production is a gross estimation because other significant factors, such as the rainfall, influence the mine water inflow. In many cases, water inflow increases more with the extension of the mine than—for a given extension—with depth.

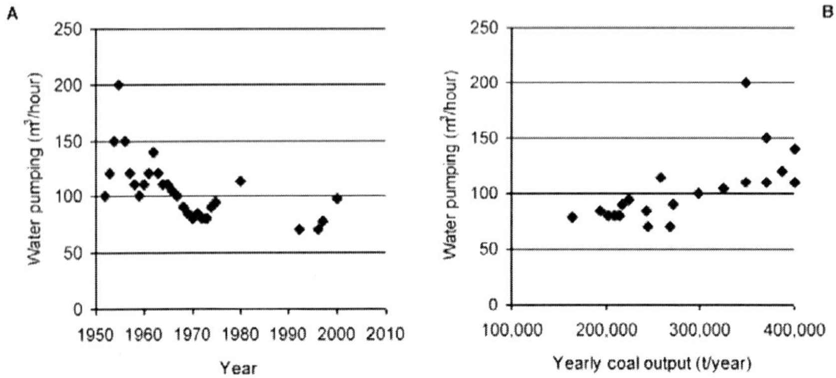


**Figure 3.** **(A)** Water pumped at La Camocha colliery per year; **(B)** Water pumped at La Camocha colliery *versus* annual coal output.

The second factor required is the air flow rate in the mine. This parameter does not vary greatly year by year; so, to develop a simple procedure, a unique representative value will be chosen. Thus, the average quantity of air flow extracted from mine by the main fans determined from measurements taken during the last years of the history of the mine is a representative value. Here, air flow rate supplied to the mine by two main exhaust fans was aproximately 60 $m^3/s$, and was practically constant over the years. When the value of this parameter is not known, it can be deduced from mining experience. Figure 4A,B represent the specific methane emissions in $m^3$ of methane per raw ton and the average methane flow in $m^3$ of gas per day for a typical underground mine in Asturias [26,27,28]. In this mine, the average output in the last 50 years is 300,000 tonnes of saleable coal per year (about 2500 of raw tonnes per day). This mine was not very gassy so, for this output level, the methane flow is of about 10,000 $m^3$ of gas per day or 0.115 $m^3/s$. Methane concentration in ventilation return in our coal mines is usually about 0.20%; consequently, the fresh airflow rate is about 60 $m^3/s$.

The last factor is electrical power. Traditionally, these mines have low mechanisation. We can assume that electrical power of mining equipment in the mine is lower than E = 1.0 MW.

It now becomes easy to estimate $k_{min}$. The maximum value of recorded $Q_w$ has to be selected. In this case, for 200 $m^3/h$, $Q_w$= 0.055 $m^3/s$. On the other hand, $Q_a$ = 60 $m^3/s$.

By replacing these values in Equation (9):

$$W_{tmin} = 25.1 \times Q_w + 0.013 \times Q_a - 0.10 \times E = 1.38 + 0.78 - 0.10 \approx 2.1 \text{ MWt} \quad (11)$$

This is a real value for geothermal power that can be extracted from La Camocha mine and, although $Q_w$is a maximum, $W_{tmin}$ is a minimum because the water inflow in this mine can be considered low.

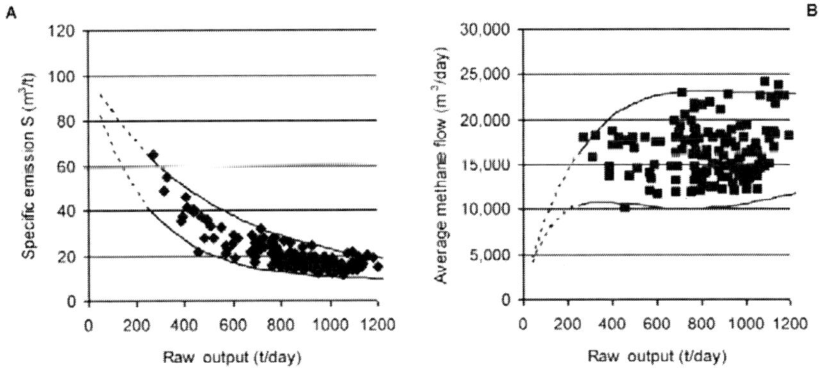


**Figure 4.** Typical methane emissions parameters from underground coal mines in Asturias (**A**) Specific emissions; (**B**) Average methane flow.

In order to estimate a geothermal power for a given total accumulated coal output, the relationship between $W_{tmin}$ and $P_T$ is calculated for all the years for which $P_T$ is known (Figure 5A and Equation (10)):

$$k_{min}(t) = W_{tmin}/P_T(t) = 2.1/P_T(t) \tag{12}$$

In order to better understand the results, this relationship is represented against the accumulated output $P_T$ in Figure 5B.

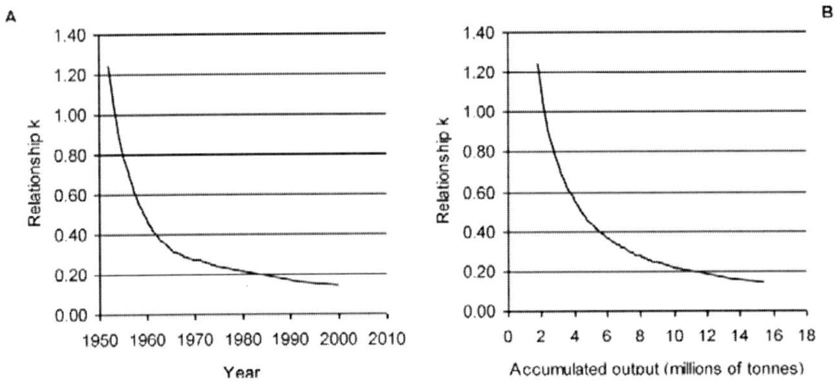


**Figure 5.** (**A**) Relationship between geothermal power and output per year (La Camocha colliery); (**B**) Relationship between geothermal power and output *versus* accumulated output (La Camocha colliery).

## 3.2. Case History 2: Figaredo Colliery

The second example is similar to the previous one in some aspects; for example, in the latter also vertical coal seams were also mined. However, there are several factors which significantly influence the water inflow into the mine, making it greater than in the former case. One factor is that the longwall method with caving was used more extensively thereby increasing the permeability of the rockmass. The second aspect is that there is a river above the mine and it has been demonstrated that a stream of water had flown from the river into the mine. The last issue is that this mine is connected to three other collieries and could receive water from them. For all these reasons, this practical case will be used to estimate a maximum value of k.

The history of coal production in this colliery is shown in Figure 6. The exploitation of coal started at the end of the XIX century (Figure 6A). It had increased during the second half of the 20th century due to the Second World War and the subsequent petroleum crisis. Peak production was reached in the decade of the 80s and afterwards production felt until the closure of the mine in 2007. Total accumulated production was about 10 million tonnes of saleable coal over more than 100 years (Figure 6B).

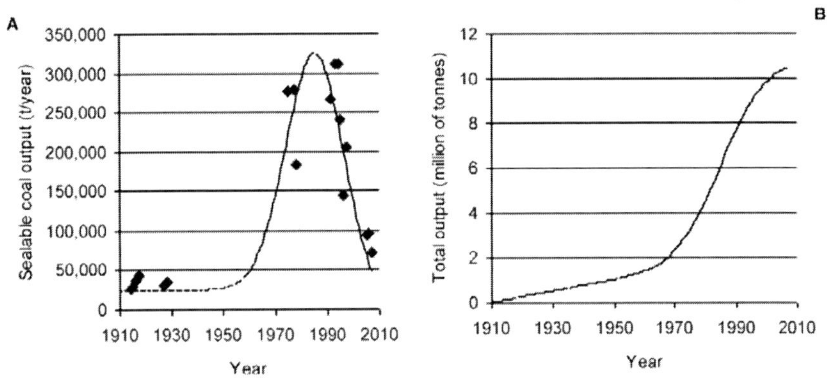


**Figure 6.** Coal production at Figaredo colliery (Asturias, Spain) (**A**) Coal output, in t/year; (**B**) Cumulative total output from 1910 to mine closure.

Figure 7A shows the water pumped from the mine for 20 years and, in Figure 7B, water flow rate pumped from the mine is related to the coal output. As can be inferred from Figure 7, water inflow in this mine is greater than in the mine previously analysed, in particular it is as much as five times higher. The maximum quantity of pumped water reached 1000 $m^3/h$ or 0.277 $m^3/s$. On the other hand, the air flow rate recorded in the last year of the mine was 90 $m^3/s$ and the electrical power of mining equipment is lower than E = 1.0 MW.

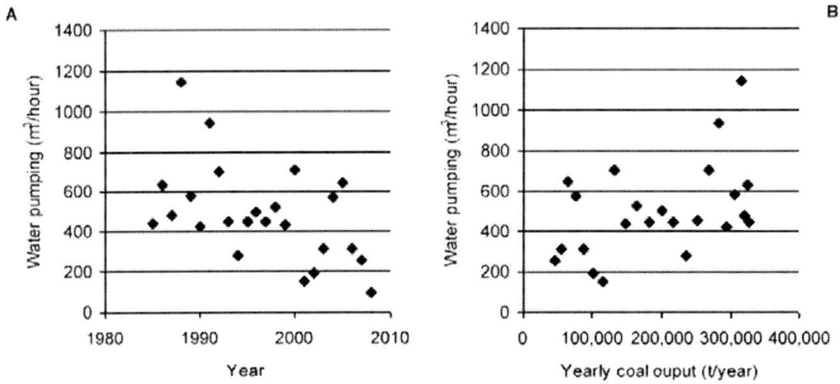


**Figure 7.** (**A**) Water pumped at Figaredo colliery per year; (**B**) Water pumped at Figaredo colliery *versus* yearly coal output.

Proceeding as in the previous case, geothermal power could be obtained from Equation (9):

$$W_{tmax} = 25.1 \times Q_w + 0.013 \times Q_a - 0.10 \times E = 6.97 + 1.17 - 0.10 = 8.1 \text{ MWt} \quad (13)$$

This is a real value for geothermal power which can be extracted from the mine and it can be considered as a maximum by comparing it with other mines in Asturias. In order to estimate geothermal power for a given total accumulated coal output, the relationship $W_{tma}/P_T$ is calculated for all the years in which $P_T$ is known (Figure 8A and Equation (10)):

$$k_{max}(t) = W_{tmax}/P_T(t) = 8.1/P_T(t) \quad (14)$$

This relationship is represented as a function of the accumulated output $P_T$ in Figure 8B, in order to better understand the results.

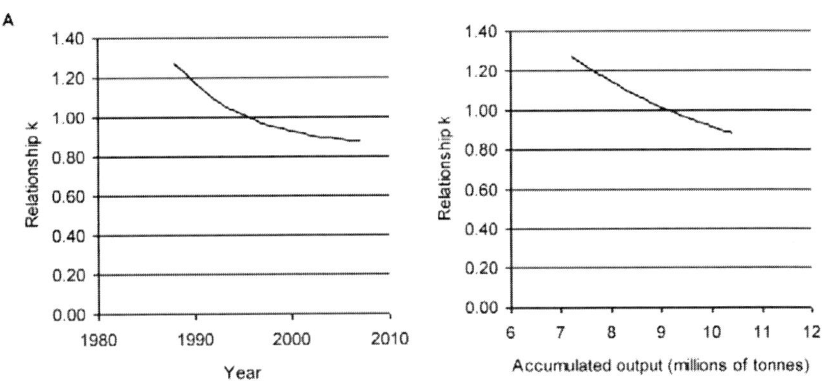


**Figure 8.** (**A**) Relationship between geothermal power and output per year (Figaredo colliery); (**B**) Relationship between geothermal power and output *versus* accumulated output (Figaredo colliery).

## 3.3. Determination of Parameter k Based on Experience

In Figure 9, the value of k has been represented as a function of the accumulated coal output for the period of activity of these two representative mines. This method for selecting the proper value of k would be useful for mines producing, at least, 5 million tonnes of saleable coal.

As it can be deduced directly from experience (Figure 9), a reasonable minimum value for k would be $k_{min} = 0.2 - 0.4$ while, a reasonable maximum for the parameter would be $k_{max} = 0.90 - 1.20$.

This means that, with regard to the assessing geothermal power in Asturias, the value $k_{min}$ would be a conservative value and this geothermal power could be actually reached, while $k_{max}$ is an optimistic value and it would be hard (even impossible) to reach the corresponding estimated geothermal power.

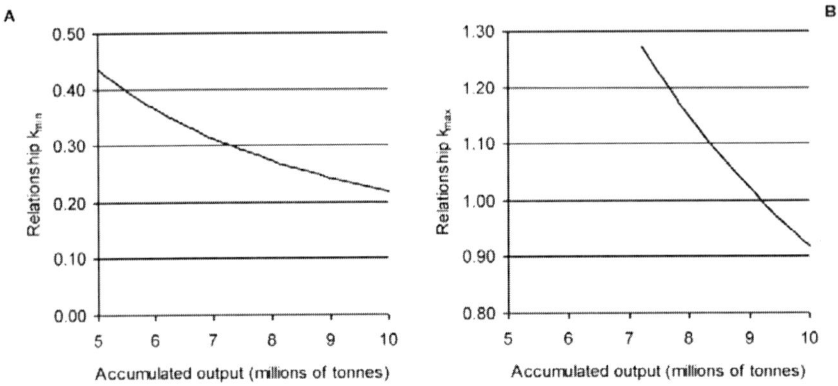


**Figure 9.** **(A)** Minimum relationship between geothermal power and output *versus* accumulated output; **(B)** Maximum relationship between geothermal power and output *versus* accumulated output.

## 4. ANALYTICAL AND SEMI-EMPIRICAL ESTIMATION OF LIMIT VALUES FOR PARAMETER K

It is not easy to estimate "theoretically" reasonable maximum and minimum values for geothermal power. But a semi-empirical approach could be used if the total length of galleries in the mine is known. If this length parameter is unknown it could be either obtained from historical data or it could also be deduced from coal output as explained below.

After a study carried out in Spain in 1990 [29], the length of galleries excavated in rock in Spanish coal mines varies from 4 to 9.5 km per million of saleable tonnes. In the case of gateroads or galleries advanced in coal seams this value ranges from 6 to 12 km. The report gave data from a total of about 25 underground mines in Spain (pit-coal, anthracite and lignite). In this period, a number of large-scale mines were in operation and these values can be taken as representative for any mine. In the following, a minimum value for the gallery length excavated in rock and a maximum value for the total length of excavated galleries (rock + coal) will be necessary. Taking into account the above, limits of 5 km and 20 km per million of tones have been chosen.

Figure 10 shows the total yearly length of galleries excavated in rock and gateroads excavated in coal seams for several mines in Asturias. The value is related to the yearly coal output and it is given in mm per ton (equivalent to kilometre per million tonnes). These Figures illustrate that these values, 5 and 20 km, could be accepted for mines in Asturias rather than show data from which these values could be deduced mathematically.

In order to define a procedure, a typical mine in Asturias with a total output of about 10 million of saleable coal has been assumed. The total necessary gallery length would be about 200 km. This value is in agreement with real data, since the total length of galleries excavated in Figaredo Colliery has been about 254 km.

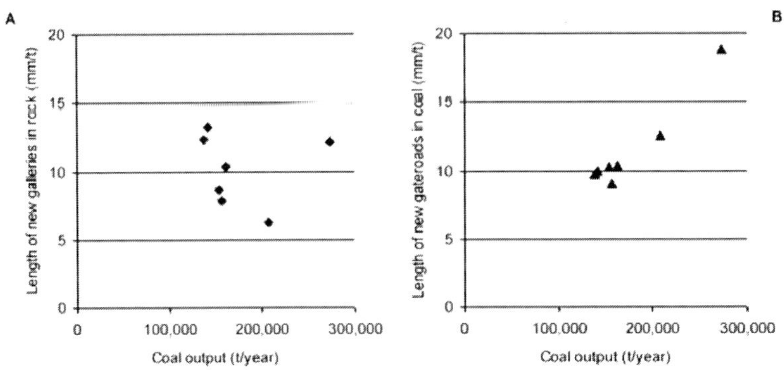


**Figure 10.** (**A**) Yearly ratio of length of new galleries excavated to coal output; (**B**) Yearly ratio of length of new gateroads excavated to coal output.

After research in a typical mine in Asturias [30], it has been found that the potential geothermal power of a 1 km gallery is approximately 50 kWt = 0.050 MWt. This means that the total geothermal power of a mine which has produced $P_T$ tonnes along its life would be:

$$W_{tmax} = 0.050 \times 20 \times P_T = 1.0 \times P_T \quad (15)$$

Therefore the value of k which can be taken as a maximum would be a constant:

$$k_{max}(t) = W_{tmax}/P_T(t) = 1.0 \quad (16)$$

This value has to be considered a "maximum" because the factor 50 kW/km was deduced from data of galleries at a depth of 500 m, where rockmass temperature was about 28 °C. It is clear that galleries at a lower depth would have smaller geothermal potential. On the other hand, it is assumable that most of the galleries excavated in rock maintain their section, with no significant convergences. Another assumed factor is that the distance between galleries is enough to allow extraction of the maximum heat from the rockmass, which is not always realistic. Finally, in order to recover this amount of heat, large quantities of water should be used (which is not always possible).

A more conservative value can be estimated if it is assumed that gateroads (galleries following coal seams) would collapse and water could not flow through them. In this case, only galleries excavated in rock are stable in the long term and only these galleries could behave as paths for water flow. Moreover, the smaller ratio of galleries' length excavated in rock to coal output is chosen. Consequently, only 5 km of galleries are useful per million of coal tonnes:

$$W_{tmax} = 0.050 \times 5 \times P_T = 0.25 \times P_T \quad (17)$$

Consequently the value of k which can be taken as a minimum is constant and its value is:

$$k_{min}(t) = W_{tmin}/P_T(t) = 0.25 \qquad (18)$$

These theoretical results are represented in Figure 11, in a graphical output similar to the one that shows the more experimental results previously deduced (in Figure 9).

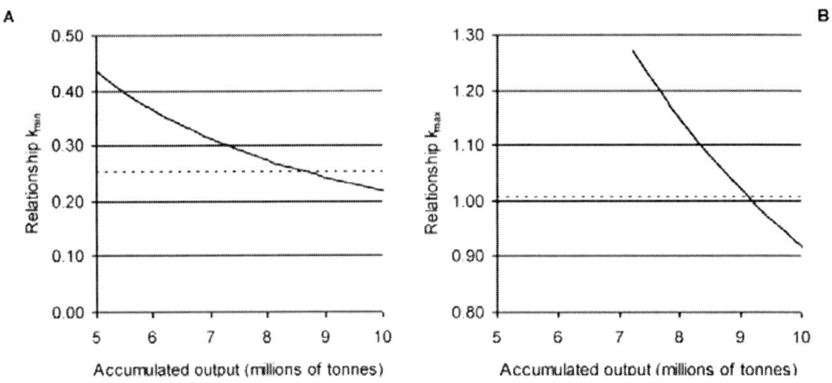


**Figure 11.** (**A**) Minimum relationship between geothermal power and output *versus* accumulated output; (**B**) Maximum relationship between geothermal power and output *versus* accumulated output.

# 5. USING THE MODEL

## 5.1. Estimation of the Geothermal Power of a Mine

It now becomes easy to estimate the geothermal power potential of any mine in Asturias, by applying the above-described method.

The case example selected to validate the method is a coal mine having mainly low, steep coal seams at a moderate depth of 400 m. The coal is anthracite, without methane and the exploitation method is longwall with caving. This coal field did not have easy access to the rest of the region, so its mining history starts about the middle of the 20th century, when a power station was built near the coalfield (Figure 12). Production drastically increased in 2000, due to the mechanisation of the works in order to mine a 4 m thick coal seam by the longwall method [31].

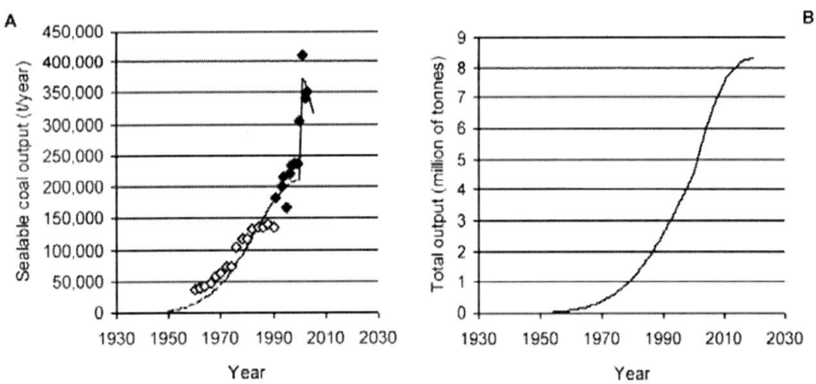


**Figure 12.** Coal production of an underground mine in Asturias (**A**) Coal output, in t/year; (**B**) Cumulative total output from approximately 1954 to present time.

Studies conducted during 2003 have provided some valuable information, which can be used to validate the approach. Up to this date, the mine had produced about $P_T = 5 \times 10^6$ tonnes of saleable coal. Taking the value $k_{min} \approx 0.25$, the minimum expected geothermal power of the mine would be:

$$W_{tmin} = 0.25 \times 5 = 1.25 \text{ MWt} \quad (19)$$

This minimum value can be verified from real data obtained from the mine. In 2003, the total water inflow into the mine was $Q_w \approx 162$ m$^3$/h = 0.045 m$^3$/s while the air flow rate was $Q_a \approx 25$ m$^3$/s [24,32]. In this case, it can also be assumed that the electrical power of mining equipment is about E = 1.0 MW.

Consequently, a realistic value for geothermal power would be:

$$W_t = 25.1 \times Q_w + 0.013 \times Q_a - 0.10 \times E = 1.13 + 0.32 - 0.10 = 1.35 \text{ MWt} \quad (20)$$

This value is greater than the previously calculated minimum value.

Taking $k_{max} \approx 1.0$, the upper limit for geothermal power (which is not expected to be reached) would thus be:

$$W_{tmax} = 1.0 \times 5 = 5.0 \text{ MWt} \quad (21)$$

Nevertheless, in opposition to estimation of the minimum parameter, the assessment of the maximum value cannot be proved.

## 5.2. Estimation the Geothermal Power for Several Mines in the Same Coalfield

This section shows the typical problem of estimating the geothermal power potential of many abandoned mines for a given coalfield in Europe, applying it to Asturian mines. The total underground coal output in Asturias during the last 200 years is shown in Figure 13. It is a fairy moderate production of only 110 million tonnes. Actually, the output is about 1 million tonnes per year. Data is only referred to pit-coal produced in the Central Coal Basin.

Actual data from 1980 and 2004 allow verification of the simplified method. In 1980, total coal ouput of mines in Asturias was about 89 million tonnes of saleable coal ($P_T = 8.9 \times 10^7$ tonnes). This production was obtained mainly from 25 collieries, so the average saleable production reached was about 3.5 Mt per mine facility. So, at the present time these mines could be considered "old mines" and the approach could be used.

For the following minimum and maximum values, $k_{min} \approx 0.25$ and $k_{max} \approx 1.0$, the minimum and maximum expected geothermal power of the mines would thus be:

$$W_{tmin} = 0.25 \times 89 = 22.2 \text{ MWt} \quad (22)$$

$$W_{tmax} = 1.00 \times 89 = 89.0 \text{ MWt} \quad (23)$$

It can be estimated that the total mine water evacuated from underground mines in Asturias was more $4.0 \times 10^7$ $m^3$ per year or $Q_w = 1.2$ $m^3/s$, as a study about mine water carried out in 1980 reports [33]. On the other hand, following the procedure explained above, coal output for this year was $5.5 \times 10^6$ tonnes, and the total air flow rate supplied to the mines would be more than 1500 $m^3/s$. With this input data, the actual geothermal power could then be estimated as:

$$W_t = 25.1 \times Q_w + 0.013 \times Q_a = 30.1 + 19.5 = 49.6 \text{ MWt} \quad (24)$$

This value is higher than the minimum geothermal power estimated before and, it is obviously lower than the maximum one. In 2004, the total coal output of mines in Asturias was about 104 million tonnes of saleable coal ($P_T = 1.04 \times 10^8$ tonnes). Taking the values $k_{min} \approx 0.25$ and $k_{max} \approx 1.0$, the minimum and maximum expected geothermal power would be:

$$W_{tmin} = 0.25 \times 104 = 26.0 \text{ MWt} \quad (25)$$

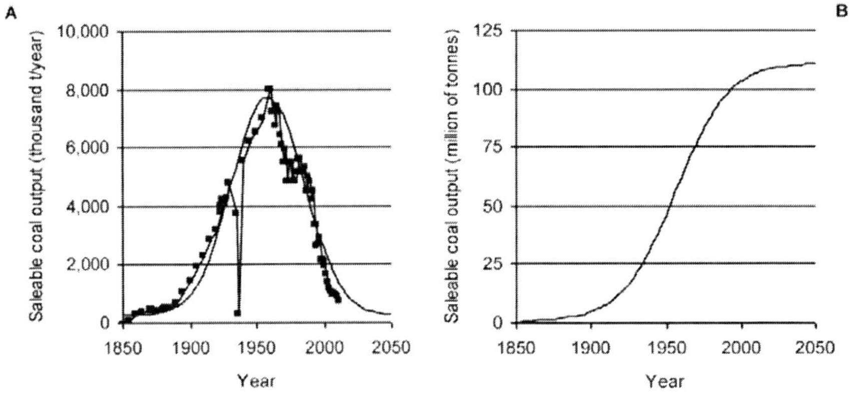


**Figure 13.** Total coal production of underground mines in Asturias (**A**) Coal output, in thousand t/year; (**B**) Cumulative total output from 1850 to present time, in million tonnes.

$$W_{tmax} = 1.00 \times 104 = 104.0 \text{ MWt} \quad (26)$$

A study carried out in 2004 [25] demonstrated that the total mine water pumped from underground mines in the Central Coal Basin in Asturias was more than $36 \times 10^6$ m$^3$ per year or $Q_w = 1.1$ m$^3$/s. On the other hand, following the procedure developed in this research, for a yearly coal output of approximately $1.8 \times 10^6$tonnes, and a total air flow rate supplied to the mines of more than 500 m$^3$/s in 2004, the estimated geothermal power is:

$$W_t = 25.1 \times Q_w + 0.013 \times Q_a = 27.6 + 6.5 = 34.1 \text{ MWt} \quad (27)$$

A value which is also between the minimum and maximum values previously estimated.

It is important to point out that, as recorded in [25], the population of villages and towns close to these mines reaches 500,000 inhabitants which could directly use this geothermal power.

## 5.3. Could the Total Geothermal Power of Abandoned Mines in Europe be Estimated?

An accurate quantification of the geothermal power of abandoned mines in Europe would not only contribute to making the right decisions but also help to find proper uses for existing funds. This is obviously an interesting problem which cannot be solved at this stage of research; the main but not the only reason, is that at the present time, thousands of mines remain abandoned in Europe with no information available and without reported data; nevertheless, at least, an attempt to estimate the potential of abandoned coal mines could be made by applying the proposed method.

The graphical output in Figure 14A shows a gross estimation of the total coal production in the European Union for the last 150 years [34,35]. The accumulated coal output could reach the value of 11,000 million tonnes (Figure 14B).

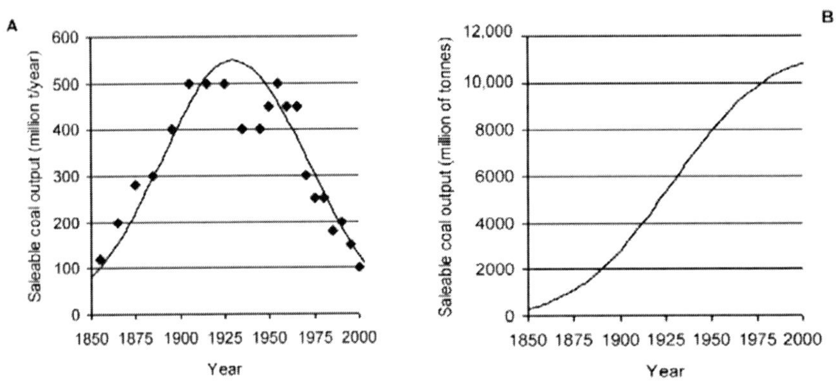


**Figure 14.** Estimation of total coal production of underground mines in Europe (**A**) Coal output, in million t/year; (**B**) Cumulative total output from 1850 to 2000, in million tonnes.

Taking the value $k_{min} \approx 0.25$, the total geothermal power potential in Europe could be assessed:

$$W_{tmin} \approx 0.25 \times P_T = 0.25 \times 11000 = 2750 \text{ MWt} \quad (28)$$

This is, about 3000 MWt could be extracted only from abandoned underground coal mines. The conclusion is that, as stated above, a vast geothermal potential from abandoned mines is not being exploited in Europe.

Furthermore, at this point, it would be interesting to make a gross estimation of hypothetical reduction in $CO_2$ emissions due to the use of this unexploited geothermal power.

Assuming that $W_t$ = 3000 MWt, and for a coefficient of performance COP = 4, the useful thermal power is:

$$W_u = W_t \times COP/(COP - 1) = 3000 \times 1.33 \approx 4000 \text{ MW} \quad (29)$$

Assuming that the power is used h = 24 h/day and d =30 days/month during m = 6 months/year, the total energy would thus be:

$$E_u = W_u \times (12 - m) \times d \times h = 4000 \times 6 \times 30 \times 24 \approx 17{,}280{,}000 \text{ MWh/year} = 17.3 \text{ TWh/year} \quad (30)$$

The ratio between tonnes of $CO_2$ emissions and MWh produced depends on the source. In order to produce 1 MWh of thermal energy, it is necessary to emit 0.850, 0.450 or 0.200 tonnes of $CO_2$ to the atmosphere depending on wether electrical, fuel or natural gas has been used as a primary energy source [30]. Assuming an average ratio of 500 t/MWh the production of 17.3 TWh/year would imply a total emission of more than 8.5 million tonnes per year.

The value of this ratio is 0.170 for geothermal power by means of heat pumps. Consequently, in this case, $CO_2$ emissions would be only 3 million tonnes per year, thus yielding a reduction of $CO_2$ emissions of about 5 million of tonnes/year.

Finally, it is important to point out that this 3000 MWt could be extracted only from abandoned coal mines. It is unquestionable that a quantity of similar magnitude could be extracted from base-metals mines. So, the total amount of geothermal energy which could be recovered from underground mines in Europe could be as much as 6000 MWt. This value is equivalent to the energy supplied by 6000 eolic generators or equivalently, to the energy supplied by a wind power park with more than 150 generators for each country in the European Union. For this reason, promoting the widespread use of this source of renewable energy is of the most importance.

## 6. CONCLUSIONS

The following conclusions can be deduced from research carried out so far:

- Although mines present a high potential for geothermal utilization, there are only few cases known in Europe where this potential has been detected, and accurately used.
- A method has been developed to allow a non-complex estimation of the limits for the geothermal potential of an abandoned underground coal mine, from the value of its total production.
- The method is useful for making geothermal resource estimates for given mining regions where coal extraction data are available; it should not be used to design a geothermal system at a mine site.
- The specific maximum and minimum values, $k_{min}$ = 0.25 and $k_{max}$ = 1.0, could also be applied in coal regions similar to Asturias. Many parameters can influence these values, as for example thermal properties and hydrogeological characteristics of the rockmass, average temperature of virgin rock and gradient of temperature

with depth, climate and average temperatures of the air and the river water and mining methods...*etc.* Consequently, values of $k_{min}$ and $k_{max}$ could be different in other regions.

- Assuming that the application of the formula has a high level of uncertainity, it has been estimated that an underground coal mine has a geothermal power of approximately 2.5 MWt per each 10,000,000 of tonnes produced.
- At least approximately 3000 MWt could be used from underground coal mines in the European Union, without including base-metals mines; the potential for coal mines is equivalent to 3,000 eolic generators or thereabouts, to the energy supplied by a wind power park with 90 generators for each country in the European Union.
- If this energy potential were used, an important reduction in $CO_2$ emissions of approximately 5 million tonnes of $CO_2$ per year could be reached.
- A good practice in mining management would be to make some mine-measurements, such as recording air flow rates, quantity of water actually pumped or air and water temperatures; this data would be of the most interest for future studies, especially when approaching the mine closure date.

## AUTHOR CONTRIBUTIONS

Rafael Rodríguez conceived and developed the idea behind the present research and he proposed the methodology and analytic procedure. Rafael Rodriguez and María B. Díaz have carried out the literature review and manuscript preparation including the searching and study of historical data of mines used in the definition of $k_{min}$ and $k_{max}$. Final review, including final manuscript corrections, was done by Rafael Rodríguez.

## REFERENCES

1. Hall, A.; Scott, J.A.; Shanga, H. Geothermal energy recovery from underground mines. *Renew. Sustain. Energy Rev.* **2011**, *15*, 916–924.

2. Klinger, C.; Charmoille, A.; Bueno, J.; Gzyl, G.; Garzon Súcar, B. Strategies for follow-up care and utilisation of closing and flooding in European hard coal mining areas. *Int. J. Coal Geol.* **2012**, *89*, 51–61.

3. Younger, P.L. Hydrogeological challenges in a low-carbon economy. *Q. J. Eng. Geol. Hydrogeol.* **2014**,*47*, 7–27.

4. Jessop, A.M.; MacDonald, J.K.; Spence, H. Clean energy from abandoned mines at Springhill, Nova Scotia. *Energy Sources* **1995**, *17*, 93–106.

5. Jessop, A.M. Geothermal energy from old mines at Springhill, Nova Scotia, Canada. In Proceedings of the World Geothermal Congress, Florence, Italy, 18–31 May 1995; pp. 463–468.

6. Raymond, J.; Therrien, R. Low-temperature geothermal potential of the flooded Gaspé Mines, Québec, Canada. *Geothermics* **2008**, *37*, 189–210.

7. Ghoreishi, S.A.; Ghomshei, M.M.; Hassani, P.; Abbasy, F. Sustainable heat extraction from abandoned mine tunnels: A numerical model. *J. Renew. Sustain. Energy* **2012**, *4*.

8. Grasby, S.E.; Allen, D.M.; Bell, S.; Chen, Z.; Ferguson, G.; Jessop, A.; Kelman, M.; Ko, M.; Majorowicz, J.; Moore, M.; *et al. Geothermal Energy Resource Potential of Canada*; Open File 6914; Geological Survey of Canada: Ottawa, ON, Canada, 2012.

9. Renz, A.; Ruhaak, W.; Schatzl, P.; Diersch, H.-J.G. Numerical modelling of geothermal use of mine water: Challenges and examples. *Mine Water Environ.* **2009**, *28*, 2–14.

10. Raymond, J.; Therrien, R. Optimizing the design of a geothermal district heating and cooling system located at a flooded mine in Canada. *Hydrogeol. J.* **2014**, *22*, 217–231.

11. Malolepszy, Z. Man-made, low-temperature geothermal reservoirs in abandoned workings of underground mines on example of Nowa Ruda coal mine, Poland. In Proceedings of the International Geothermal Conference, Reykjavík, Iceland, 14–17 September 2003; pp. 23–29.

12. Malolepszy, Z. Low temperature, man-made geothermal reservoirs in abandoned workings of underground mines. In Proceedings of the 28th Workshop on Geothermal Reservoir Engineering, Stanford, CA, USA, 27–29 January 2003; Stanford University: Stanford, CA, USA, 2003; pp. 259–265.

13. Malolepszy, Z.; Demollin-Schneiders, E.; Bowers, D. Potential use of geothermal mine waters in Europe. In Proceedings of the World Geothermal Congress 2005, Antalya, Turkey, 24–29 April 2005; International Geothermal Association: Antalya, Turkey, 2005; pp. 1–3.

14. Demollin-Schneiders, E.; Malolepszy, Z.; Bowers, D. Potential use of geothermal energy from mine water in Europe for cooling and heating. In Proceedings of the International Conference Passive and Low Energy Cooling for the Built Environment, Santorini, Greece, 19–21 May 2005; Santamouris, M., Ed.; Heliotopos Conferences: Santorini, Greece, 2005; pp. 683–685.

15. Tóth, A.; Bobok, E. A prospect geothermal potential of an abandoned copper mine. In Proceedings of the 32nd Workshop on

Geothermal Reservoir Engineering, Stanford, CA, USA, 22–24 January 2007; pp. 1–3.

16. Hamm, V.; Bazargan Sabet, B. Modelling of fluid flow and heat transfer to assess the geothermal potential of a flooded coal mine in Lorraine, France. *Geothermics* **2010**, *39*, 177–186.

17. Ferket, H.L.W.; Laenen, B.J.M.; Van Tongeren, P.C.H. Transforming flooded coal mines to large-scale geothermal and heat storage reservoirs: What can we expect? In Mine Water—Managing the Challenges, Proceedings of the International Mine Water Association Congress 2011, Aachen, Germany, 4–11 September 2011; Rüde, T.R., Freund, A., Wolkersdorfer, C., Eds.; International Mine Water Association: Wendelstein, Germany, 2011; pp. 171–176.

18. Uhlík, J.; Baier, J. Model evaluation of thermal energy potential of hydrogeological structures with flooded mines. *Mine Water Environ.* **2012**, *31*, 179–191.

19. *Municipal building, Park Hills, Missouri*; Ghpc #CS-064; Geothermal Heat Pump Consortium: Washington, DC, USA, 1997.

20. Watzlaf, G.R.; Ackman, T.E. Underground mine water for heating and cooling using geothermal heat pump systems. *Mine Water Environ.* **2006**, *25*, 1–14.

21. Guo, P.; He, M.; Yang, Q.; Chen, C. Wellhead anti-frost technology using deep mine geothermal energy. *Min. Sci. Technol. China* **2011**, *21*, 525–530.

22. He, M.; Zhang, Y.; Guo, D.; Qian, Z. Numerical analysis of doublet wells for cold energy storage on heat damage treatment in deep mines. *J. China Univ. Min. Technol.* **2006**, *16*, 278–282.

23. Zhang, Y.; Guo, D. Effect of cold energy storage of doublet-wells aquifer thermal energy storage in Sanhejian coal mine. *Energy Procedia* **2012**, *14*, 1730–1734.

24. Toraño, J.; Rodríguez, R.; Rivas, J.M. Application of numerical methods in the analysis of the hydrogeology of an area affected by underground mining. In Proceedings of the Congresso de Métodos Computacionais em Engenharia, Lisboa, Portugal, 31 May–2 June 2004; Asociación Portuguesa de Mecánica Teórica Aplicada y Computacional (AMPTAC). Servicio de publicaciones del CIMNE: Barcelona, Spain, 2004; pp. 1–15.

25. Jardón, J.; Pendas, F.; Ordóñez, A.; Cordero, C.; Álvarez, C.; Garzón, B. Aprovechamiento de las aguas de mina en la Cuenca Central Asturiana como recurso hídrico y energético. In Proceedings of the XII International Congress on Energy and Mineral Resources, Oviedo, Spain, 7–11 October 2007; Consejo Superior de Colegios de Ingenieros de Minas: Madrid, Spain, 2007; pp. 1–7. (In Spanish).

26. Luque, V.C.; Pedeca, S.C.L.E. *Manual de ventilación de minas*; Pedeca S. Coop. Ltda: Madrid, Spain, 1988; p. 732. (In Spanish)

27. Díaz Aguado, M.B.; González Nicieza, C. Control and prevention of gas outbursts in coal mines, Riosa-Olloniego coalfield, Spain. *Int. J. Coal Geol.* **2007**, *69*, 253–266.

28. Rodríguez, R.; Lombardía, C. Analysis of methane emissions in a tunnel excavated through Carboniferous strata based on underground coal mining experience. *Tunn. Undergr. Space Technol.* **2010**, *25*, 456–468.

29. Rambaud, C. Situación actual y tendencias futuras en las labores de preparación. In Proceedings of the Jornadas Técnicas sobre Labores de Preparación, Madrid, Spain, 21–22 November 1990;

Instituto Tecnológico Geominero de España: Madrid, Spain, 1990; pp. 23–25. (In Spanish).

30. Rodríguez, R.; Díaz, M.B. Analysis of the utilization of mine galleries as geothermal heat exchangers by means a semi-empirical prediction method. *Renew. Energy* **2009**, *34*, 1716–1725.

31. Alvarez, T.J.; Rodríguez, R.; Rivas, J.M.; Casal, M.D. Economic and technical results mining a 4 m thick coal seam in Spanish Carbonar Colliery. *Glückauf* **2013**, *139*, 323–328.

32. Toraño, J.; Rodríguez, R.; Rivas, J.M. New techniques *versus* experience: A real case of mine ventilation analysis. In Proceedings of the 10th U.S./North American mine ventilation symposium, Anchorage, AK, USA, 16–19 May 2004; Bandopadhyay, S., Ganguli, R., Eds.; Taylor & Francis: London, UK, 2004; pp. 487–492.

33. Pendás, F.; García, M.P. Caracterización hidrogeológica de la minería de la Cuenca Central Asturiana. In Proceedings of the VIII Congreso Internacional de Minería y Metalurgia, Oviedo, Spain, 16–22 October 1988; Asociación Nacional de Ingenieros de Minas de España: Oviedo, Spain, 1988; pp. 279–298. (In Spanish).

34. Del Rosal, I. La productividad en la minería española del carbón. *Revista de Minas* **1999**, *19-20*, 213–218. (In Spanish). Del Rosal, I. La reconversión del carbón, una dependencia plena de la decisión pública. *La Empresa Pública* **2000**, *1*, 156–166. (In Spanish).

# CHAPTER 9

# Failure Mechanism Analysis and Support Design for Deep Composite Soft Rock Roadway: A Caase Study of the Yangcheng Coal Mine in China

**Bangyou Jiang,[1] Lianguo Wang,[1] Yinlong Lu,**
**[1] Shitan Gu,[2] and Xiaokang Sun[1]**

*[1]State Key Laboratory for Geomechanics and Deep Underground Engineering, China University of Mining and Technology, Xuzhou 221116, China*
*[2]College of Mining and Safety Engineering, Shandong University of Science and Technology, Qingdao 266590, China*

## ABSTRACT

This paper presented a case study of the failure mechanisms and support design for deep composite soft rock roadway in the Yangcheng Coal Mine of China. Many experiments and field tests were performed to reveal the failure mechanisms of the roadway. It was found that the surrounding rock of the roadway was HJS complex soft rock that was characterized by poor rock quality, widespread development of joint fissures, and an unstable creep property. The major horizontal stress, which was almost perpendicular to the roadway, was 1.59 times larger than the vertical stress. The weak surrounding rock and high tectonic stress were the main internal causes of roadway instabilities, and the inadequate support was the external cause. Based on the failure mechanism, a new support design was proposed that consisted of bolting, cable, metal mesh, shotcrete, and grouting. A field experiment using the new design was performed in a roadway section approximately 100 m long. Detailed deformation monitoring was conducted in the experimental roadway sections and sections of the previous roadway. The monitoring results showed that deformations of the roadway with the new support design were reduced by 85–90% compared with those of the old design. This successful case provides an important reference for similar soft rock roadway projects.

## 1. INTRODUCTION

Support for a soft rock roadway, especially for deep high-stressed composite soft rock roadways, is a worldwide problem [1]. Composite soft rock is characterized by a broken-rock structure, low rock strength, intense water swelling, and rheological response [2, 3]. In a complicated geomechanical environment (i.e., high in-situ stress, high ground temperature, high osmotic pressure, and strong mining disturbance), many roadways in deep coal mines with weak surrounding rocks experience severe deformations and failure, which not only reduces the functionality of the roadway but also endangers the safety of the mine personnel.

Recently, numerous studies have been performed on the failure mechanisms of deep soft rock roadways. He et al. performed an experiment and field investigation of the roadway of a loaded-car line that was seriously destroyed in the Xing'an Coal Mine, China, and found that the large buried depth and strong tectonic stress caused its failure [4]. Deep soft rocks are often cut by various weak planes (e.g., joint and bedding) [5]. Under high in-situ stresses, a shear slip will appear along these weak planes, which could loosen the structure of the surrounding rock and eventually caused an unstable failure of the roadway [6–8]. Seedsman and Yuan discussed the failure mechanisms of deep soft rock roadways [9, 10]. Their studies showed that low rock strength and broken-rock structure lead to huge plastic deformations and finally failure. Wu et al. obtained the same conclusion [11]. In addition, Huang et al. presented the idea that, in a high-stress environment, the soft rock had a strong rheological response and the rheological zone persistently developed toward the interior of the rock surrounding the roadway, which lead to instability [12]. However, different deep soft rock roadways have different failure mechanisms because of the complicated, changing engineering geological conditions. Based on a survey of a large number of documents, Shen concluded that various roadway failure mechanisms could be classified into six types: beam failure, joint controlled rock falls, roof sag, guttering and shear failure, skin failure, and rib failure [1]. This classification method has great value for applications and instructions. In general, a complicated geomechanical environment and weak mechanical properties of the surrounding rock are the original external and internal causes, respectively, of a deep soft rock roadway failure.

Research on the failure mechanisms of a roadway is a prerequisite of realizing the ultimate goal of a reasonable support design. Currently, bolt, anchor cable, grouting, yieldable steel sets, and combined supporting systems are widely employed in soft rock roadways in deep coal mines, but the most optimal supporting system is arguable. Bolting and bolting combined with metal net and shotcrete are extensively used in many countries such as the United States of America and Australia, whereas, in Germany, Poland, and other European countries, U-shaped steel sets are commonly used as the main support methods [13]. Because of the reasonable cross section shape and high strength, yieldable U-shaped steel supports are always used to maintain the stability of roadways in loose and broken-rock strata [14]. However, the high cost and complex process of installing these supports are the main disadvantages and fatal flaws of this support type. Recently, the yielding bolts support has become very popular. This method works by adding a yielding tube to the bolts [15]. Although this yielding support can adapt well to the deformation and pressure of a soft rock roadway, its strength and yielding capacity are limited. The yielding bolts support is seldom used in a large deformation broken soft rock roadway [16]. Wang et al. proposed a combined bolt and grouting support to control large deformations of a soft rock roadway and achieved good control of the deformations [17]. A new and effective support technology, the bolt-grouting support, has developed rapidly and has been widely employed in soft rock roadways and tunnel engineering [18–20]. Moreover, there are many other support technologies for soft rock, such as the concrete-filled steel tubular stent [21], constant resistance and large deformation bolts [22], and other combined support technologies [23–25]. However, these support technologies are unable to restrain the loose and broken rock surrounding a deep high-stress complicated soft rock roadway.

So, the failure mechanisms and support technologies of a soft rock roadway under complicated conditions have been the focus of research in mining engineering and other related fields. This paper reports a case study of the failure mechanisms and support design for a deep composite soft rock roadway in the Yangcheng Coal Mine in Shandong Province, China. In this paper, a large number of experiments and field tests were performed, and a mechanical model was developed. Furthermore, based on the failure mechanism and characteristic of the roadway, a new optimal support was proposed and more reasonable parameters were designed. The research results provide a significant reference for support design for similar roadway engineering projects.

## 2. GEOLOGY AND ENGINEERING BACKGROUND

Located in the Jibei diggings in Shandong Province, the Yangcheng Coal Mine covers a mining area of 46 $km^2$(2.2~6.7 km wide from east to west and 10.5 km long from north to south) with 153 million tons of recoverable reserves and an estimated production capacity of 2.4 million t/a. The −650 main roadway in the south wing was the horizontal development roadway that was responsible for ventilation and transportation to mining areas in the south wing for the next 12 years of production. Obviously, this roadway was very important to the whole coal mine. Due to faults, the roadway crossed over No. 3 coal seam and other weak strata on the roof and floor (Figure 1), which extremely reduced the strength and integrity of the surrounding rock.

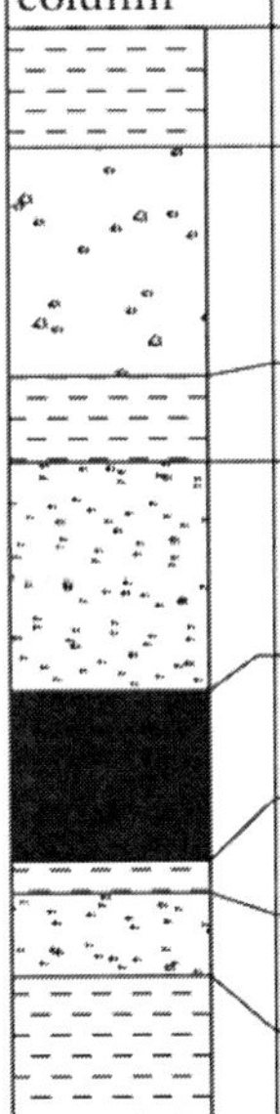

| Lithological geological column | Thickness (m) | Lithology | Lithology description |
|---|---|---|---|
| | 5.43 | Mudstone | Contained No. 2 coal seam, the thickness of which is 0.45 m |
| | 10.51 | Medium-grained sandstone | Shale cementation, existing carbonaceous texture, bedding development |
| | 4.76 | Mudstone | Gray mudstone |
| | 10.59 | Fine sandstone | Fine sandstone and siltstone interbedded, wavy bedding development |
| | 7.78 | No. 3 coal seam | Massive structure, joint development |
| | 0.6 | Mudstone | Massive structure, common slip surface |
| | 2.7 | Fine sandstone | Fine sandstone and siltstone interbedded |
| | 6.8 | Mudstone | Massive structure, silty |

**Figure 1:** Strata histogram of 3# coal seam roof and floor.

The buried depth of the main roadway was approximately 700 m. The cross section was a semicircle arch with straight walls. The walls were 1.8 m high, the arch was 2.4 m high, and the net width of the roadway was 4.8 m. Primarily, bolts combined with anchor cables and metal mesh were used as the permanent support to control the surrounding rock. Three months after excavation, the roadway is seriously distorted, and the cross section area of some sections shrank by more than 80% (Figure 2). Subsequently, the damaged roadway had been repaired and reinforced. Based on the original support, seven grouting bolts with a diameter of 26 mm and a length of 2.5 m were evenly installed in the whole cross section at an interval of 2400 mm along the roadway axis. U-shaped steel sheds with a width of 4800 mm and a height of 4765 mm were placed at an interval of 1200 mm. Furthermore, an inverted arch that matched the U-shaped steel sheds was used, and concrete was poured to 525 mm thickness. The support design is presented in Figure 3. However, the main roadway still suffered large deformations (Figure 4). Repeated repairs and reinforcements lead to the vicious circle of "excavation-repair-failure." Therefore, it is urgent to thoroughly study the failure mechanisms and optimize a support design.

the failure mechanisms and optimize a support design.

**Figure 2:** Large deformation of the roadway.

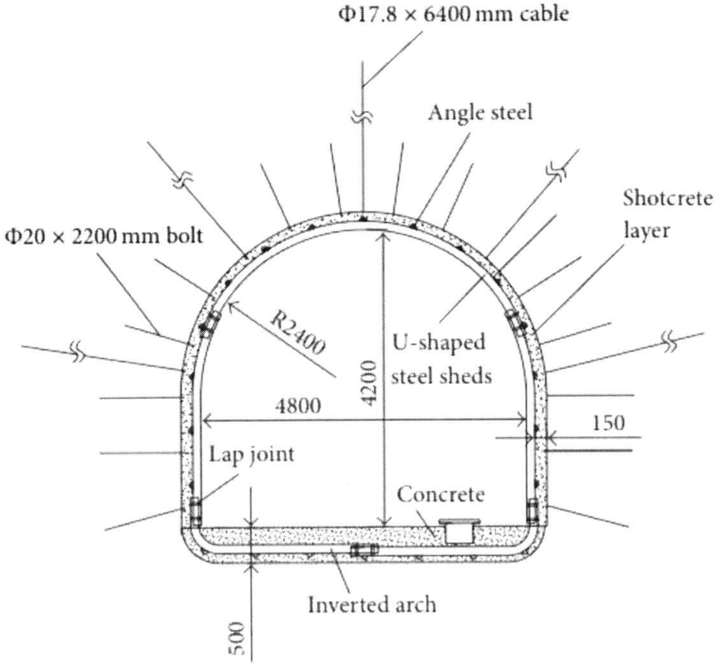


**Figure 3:** The old supporting design of the roadway.

**Figure 4:** Deformation and failure of the roadway after repairing.

# 3. EXPERIMENTS AND FIELD TESTS

## 3.1. Physical and Mechanical Properties

The rock quality designation (RQD) is an index that provides a quantitative judgment of the rock mass quality, which can be obtained from drill cores [26, 27]. Core drilling was performed in four representative locations (#1, #2, #3, and #4) in the main roadway, as shown in Figure 5. The RQD of the four locations were 21.9%, 21.4%, 43.1%, and 25.8%, respectively. The descriptive statistics of the RQD samples are presented in Table 1. From the statistics in Table 1, the rock quality at locations #1 and #2 was very poor (Classification V). Similarly, the rock quality at location #3 and #4 was poor (Classification VI). This indicated that the integrity of the rock surrounding the main roadway was low, which allowed wide development of joint fissures.

**Table 1:** Classification criterion of rock quality by the RQD [26, 27].

| RQD (%) | Classification | Rock quality |
|---|---|---|
| <25 | V | Very poor |
| 25–50 | IV | Poor |
| 50–75 | III | Fine |
| 75–90 | II | Good |
| 90–100 | I | Excellent |

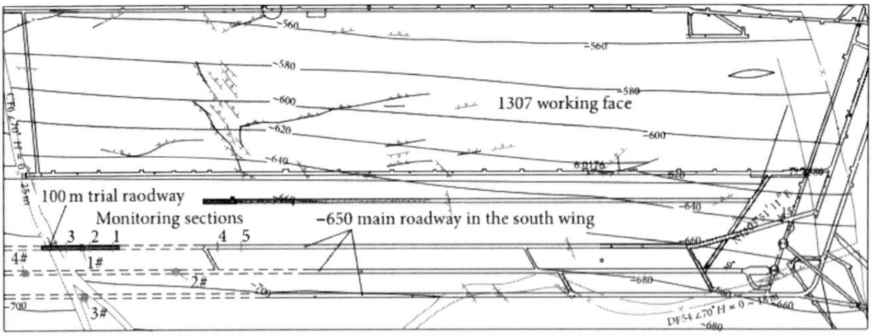


**Figure 5:** The locations of boreholes 1#~4#.

Rock samples were divided into siltstone, fine sandstone, medium-grained sandstone, and mudstone according to their lithological characteristics. Conventional physical and mechanical tests were conducted on rock samples using a MTS 815.02 rock servo-hydraulic machine. Test results are listed in Table 2. The ④ and ⑤ mudstone samples are very weak. They are typical soft rock samples due to their low uniaxial compressive strengths (24.13 MPa and 23.05 MPa) and tensile strengths (1.63 MPa and 2.22 MPa). The ①–③ sandstone samples have high strength with uniaxial compressive strengths greater than 30 MPa. The uniaxial compressive strength of the ① siltstone sample is 47.46 MPa. However, because of the wide development of joint fissures in the surrounding rock, the whole engineering rock mass is significantly distorted under the action of engineering forces; therefore, this sample has the characteristics of a jointed soft rock.

**Table 2:** Physical and mechanical parameters of surrounding rocks.

| | | Natural specimens | | | | | Saturated specimens | |
|---|---|---|---|---|---|---|---|---|
| Serial number | Lithology | Bulk density (KN·m$^{3}$) | Uniaxial compressive strength (MPa) | Tensile strength (MPa) | Poisson's ratio | Elastic modulus (GPa) | Uniaxial compressive strength (MPa) | Softening coefficient |
| ① | Siltstone | 25.70 | 47.46 | 5.82 | 0.25 | 5.89 | 22.78 | 0.48 |
| ② | Fine sandstone | 25.16 | 31.85 | 3.76 | 0.13 | 5.23 | 17.20 | 0.54 |
| ③ | Medium-grained sandstone | 24.61 | 30.14 | 4.49 | 0.18 | 5.23 | 22.00 | 0.73 |
| ④ | Mudstone | 24.65 | 24.13 | 1.63 | 0.11 | 3.10 | — | — |
| ⑤ | Mudstone | 25.75 | 23.05 | 2.22 | 0.10 | 4.30 | 6.45 | 0.28 |

To study the softening characteristics of water in the surrounding rock, a saturated water content test was conducted on the surrounding rock sample. The uniaxial compressive strengths of the water-saturated samples are presented in Table 2. The uniaxial compressive strength of water-saturated samples decreases significantly. The softening coefficient of the ⑤ mudstone in the water-saturated condition is 0.28. The ④ mudstone disintegrated in water (Figure 6). Rock samples contain expansible hydrophilic clay minerals that quickly absorb water molecules when the samples are soaking in water, which will destroy the microstructure, decrease the strength, and even disintegrate the rock samples.

**Figure 6:** Water-induced weakening process of mudstone .

In addition, uniaxial creep tests of the rock samples were performed using a multistep incremental loading. The loading rate of each stage was 40 N/s and the duration of each stage was approximately 24 h. Some results of the creep test are shown in Figure 7. The creep tests demonstrated that the surrounding rock in the main roadway has an unstable creep property and the critical strength of the unstable creep deformation of different rock samples ranges within 12~20 MPa. Based on the comprehensive analysis, the surrounding rock in the main roadway of the Yangcheng Coal Mine is an HJS complex soft rock with combined characteristics of a jointed soft rock, expansible soft rock, and high-stress soft rock [2].

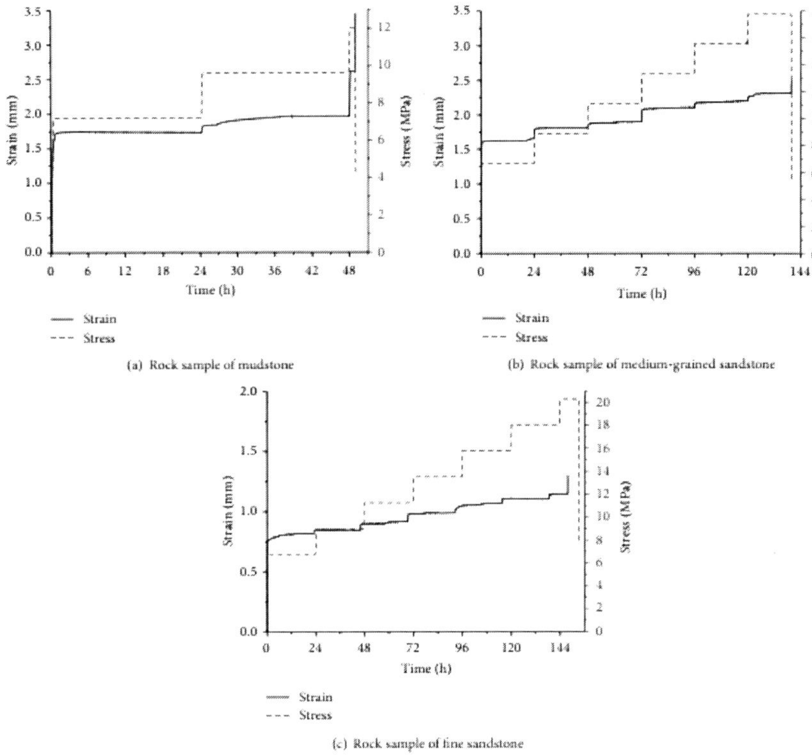


**Figure 7:** Uniaxial creep strain and creep stress curves.

## 3.2. In Situ Stresses

### 3.2.1. Gravity Stress Field

The overall buried depth of the main roadway is approximately 700 m. The mean volume-weight of the overlying rock is approximately 25 KN/m$^3$, so the gravity stress is 17.5 MPa, higher than the strength of the surrounding rock. The critical strength of the unstable creep deformation of different surrounding rock strata is 12~20 MPa, which is easily exceeded by the surrounding rock stress because of roadway excavation. This explains why the surrounding rock has large nonlinear flow deformations after roadway excavation.

### 3.2.2. Tectonic Stress Field

Based on the tectonic stress test results for the Yangcheng Coal Mine [28], the major principal stress in the in situ stress field is the horizontal stress. The azimuth angle of the maximum principal horizontal stress is 112.3°. The maximum principal horizontal stress is 27.81 MPa or 1.59 times greater than the vertical stress and 1.92 times greater than the minimum principal horizontal stress (14.47 MPa). The angle between the axis of the main roadway and maximum principal horizontal stress is 87.7°. The influence of the horizontal stress on the main roadway during excavation can lead to severe failure of the roadway.

## 3.3. Broken-Rock Zone and Failure Characteristics

The broken-rock zone and failure characteristics were detected by a borehole TV in several typical sections of the main soft rock roadway (Figure 8). The field borehole TV detection discovered that the rock surrounding the main roadway can be divided into intact, crack, and fracture zones according to the degree of failure [29], as shown in Figure 9. An intact zone is defined as a region in which no fissures are observed along a continuous distance of 30 cm of the borehole. A crack zone is defined as a region that contains 1-2 sets of fissures along a continuous distance of 30 cm of the borehole. A fracture zone is a region that contains three or more fissures or evident fracture along a continuous distance of 30 cm of the borehole.

The surrounding rock failure of the main air return roadway was analyzed based on these definitions of zones. Generally speaking, the surrounding rock is seriously distorted, with a large broken-rock zone (approximately 2~3 m and even reaching 5 m in some places) and discontinuous asymmetrical failure. Many crack zones and fracture zones are observed within a 3 m range of the surrounding rock. When the borehole depth exceeds 3 m, the crack zone and intact zone occur alternately. As the borehole goes deeper, the length of the intact zone gradually increases. The depth of the surrounding rock failure is very large, and the crack zone still can be detected at a depth of 8–10 m.

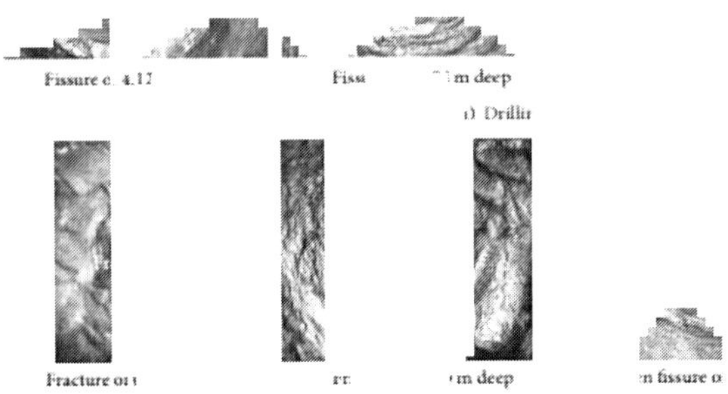

**Figure 8:** Fracture patterns of surrounding rocks from borehole TV observation.

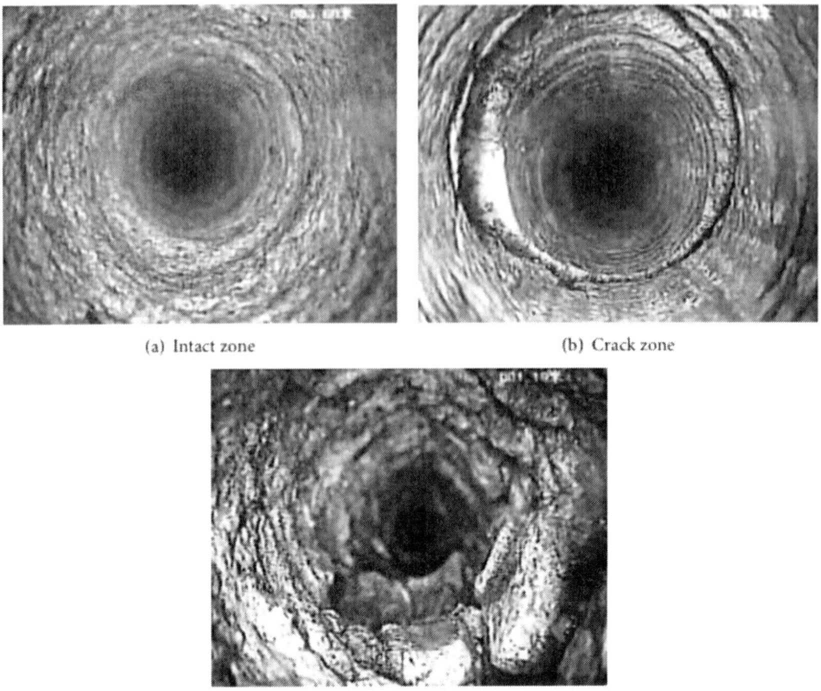

(a) Intact zone

(b) Crack zone

(c) Fracture zone

**Figure 9:** Three typical zones around the surrounding rock.

## 4. FAILURE MECHANISM OF THE MAIN SOFT ROCK ROADWAY

Experimental and field tests showed that the internal causes for the instability of the roadway were the weak surrounding rock and high tectonic stress. The unreasonable support design was the external cause.

Low strength and wide development of joint fissures of the surrounding rock significantly decrease the bearing capacity of the surrounding rock. The horizontal stress (1.59 times greater than the vertical stress), which is the major principal stress in the in situ stress field, is almost perpendicular to the roadway direction. The high horizontal stress caused a major concentration stress in the roof and floor of the roadway. Deformation and instability first occurred in the roof and floor.

After failure of the roadway roof and floor, the rock masses against the two roadway ribs had large stress concentrations in a biaxial stress state. The concentrated stress causes the abundant primary joint fissures to expand and interconnect with each other to form an unstable surrounding rock structure with a fracture zone and crack zone occurring alternately. Because the surrounding rock in the fracture zone loses bearing capacity completely, the abutment pressure of the rock surrounding the roadway migrates to the crack zone and deeper intact zone, which will turn the crack zone into a fracture zone and the intact zone into a crack zone until the surrounding rock stabilizes.

The borehole TV showed that the broken-rock zone of the roadway was approximately 2-3 m, which exceeded the effective anchorage length of the bolts. The maximum failure depth of the crack zone in the roadway extended 8–10 m, which was wider than the length of the cables. So the bolts and cables were in fact useless, and the U-shaped steel sheds were unable to support the deformations of the large quantity of broken surrounding rock.

## 5. SUPPORT DESIGN

Although the supporting capacity of the previous support was very high, the main roadway deformed seriously. The old support design is unscientific and lacking specificity. A new and better support design is needed to control the roadway deformations and instability. Based on the results of experiments and field tests, the new design should consider the following.(1)Rapid support installation after excavation [1]: the surrounding rock of the deep high-stress roadway may deform immediately after roadway excavation because of the stress concentration. Therefore, installing the bolts and cables at an early stage is essential and can avoid premature yield failure of the surrounding rock and allow the support to work with the intact rock.(2)Long cable: field tests showed that the depth of the surrounding rock failure was very large, and many cracks still can be detected 8 m deep or deeper. The cable length of the old support is 6.4 m, which is much less than the depth of the surrounding rock failure, so it cannot control the stability of the surrounding rock. The new long cable should terminate inside the relatively intact rock. So the length of new long cables should be greater than the depth of the surrounding rock failure.(3)Grouting reinforcement: field test results showed that the surrounding rock was weak because of wide development of joint fissures and a large broken-rock zone; these conditions were the main causes of the roadway instability. The best way to improve this is to use a grouting reinforcement support because this can rebuild loose and broken surrounding rock into a solid rock. Another benefit is that grouting can seal fissures of the surrounding rock and prevent infiltration of water and air, avoiding the softening effect of water and air on the rocks and increasing bearing capacity.(4)High pretension: high pretension plays an important role in roadway support because it can significantly enhance the strength of the anchorage body.

Based on these considerations, a new support design was presented. The specific layout of the bolts, cables, and so forth is shown in Figure 10. The new support combines bolting, shotcreting with wire mesh, anchor cabling, and grouting. Moreover, a deep and superficial coupling grouting support is achieved with a grouting pipe and grouting cable, as seen in Figure 10. The grouting pipe rebuilds the broken surrounding rocks by forming an integral load-bearing ring, whereas the grouting cable seals the deep fracture and stabilizes the superficial load-bearing ring. The floor is supported by an overbreak and bolt-grouting backfill technology that can increase the integrity and bending resistance of the floor. Specific support parameters in the new support design are introduced as follows.(1)Bolt parameters: there are twenty-five bolts in the roadway cross section, including eight bolts in the roof, twelve bolts in the ribs, and five bolts in the floor. The bolts have a diameter of 22 mm and a length of 2.5 m. The spacing between the bolts in the roof and ribs along the roadway axis is 0.7 m. The spacing between the bolts in the roof and ribs along the roadway radial is 0.7 m. The spacing between the bolts in the floor is 0.9 m. Metal meshes are applied before the anchor rod reinforcement. Fifty-millimeter thick shotcrete will be constructed (primary shotcrete) as soon as possible after the bolting support. The anchorage length of each bolt must be greater than 0.6 m. All bolts should be pretensioned with a pretension load of 70 kN. The bolts are made of levorotary rebar steel with a tensile strength of 500 MPa.(2)Metal mesh parameters: metal mesh is welded to steel bar that has a diameter of 6 mm. The metal mesh has a length of 2.0 m and a width of 1.0 m. The grid of the metal mesh has a length/width ratio of 0.1. The metal mesh is installed to the whole roadway section, and two adjacent metal meshes are connected by iron wire.(3)Grouting pipe parameters: there are ten grouting pipes in a roadway cross section including five in the roof, two in the ribs, and three in the floor. The grouting pipes have a diameter of 26 mm and a length of 2.5 m. The intervals between the grouting pipes along the roadway axis and radial are both 1.8 m. Grouting lines are paved immediately after the primary shotcrete support.(4)Grouting cable parameters: based on the surrounding rock deformation characteristics of the roadway measured from experiments and field tests, a new grouting cable is designed, as seen in Figure 11. The new grouting cable has many advantages such as a small grouting resistance, large flow, high efficiency, and wide application range. Gradual grouting from the deep surrounding rock to the superficial surrounding rock can be achieved with high and low pressure grouting sections of the new grouting cable. There are eleven grouting cables in a roadway cross section including five in the roof, four in the ribs, and two in the bottom angle. The grouting cables have a diameter of 22 mm and a length of 10.0 m in the roof and ribs and 6.0 m in the bottom angle. The intervals between the grouting cables in the roof and ribs along the roadway and radial axes are both 1.4 m. The spacing between the grouting cables in the bottom angle along the roadway axis is 1.4 m. The anchorage length of each grouting cable must be larger than 1.5 m. All grouting cables should be pretensioned with a pretension load of 200 kN. The grouting cables are made of steel strand with a tensile strength of 1860 MPa. Grouting cables are installed after the grouting pipes, and secondary shotcrete with the thickness of 100 mm is applied as soon as the grouting cable pavement is finished.(5)Grout parameters: the grout is a single liquid prepared from Po42.5 ordinary silicate cement with a water/cement ratio of 1 : 2. The final grouting pressure of the grouting-anchor cable is 3~4 MPa and the final grouting pressure of the grouting lines is 1.5~2 MPa [17]. The grouting is implemented within 15~18 days after roadway excavation so the grouting is most effective [30].(6)Concrete parameters: the strength of the sprayed concrete is C20 and the strength of the concrete for floor backfilling is C40, which includes an added waterproofing agent. Normal concrete easily breaks when the surrounding rock experiences a large deformation. Hence, steel fibers are added in the concrete to increase flexibility and deformability of the concrete layers. Furthermore, the metal meshes can also increase the tensile strength of the concrete layers. Backfilling the extra excavation space of the floor with concrete is conducted last.

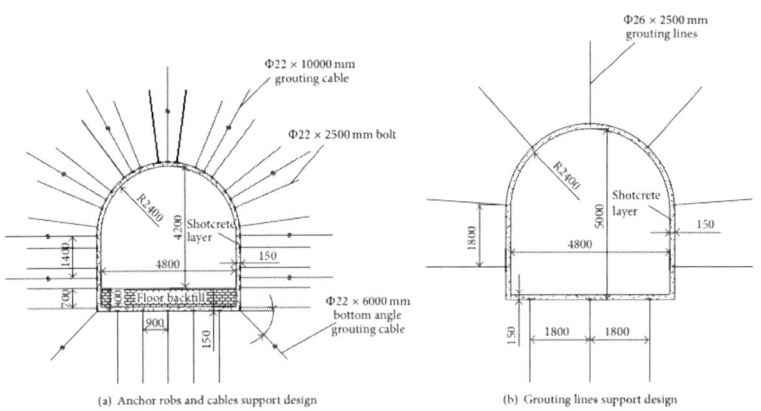


**Figure 10:** New supporting design of the roadway.

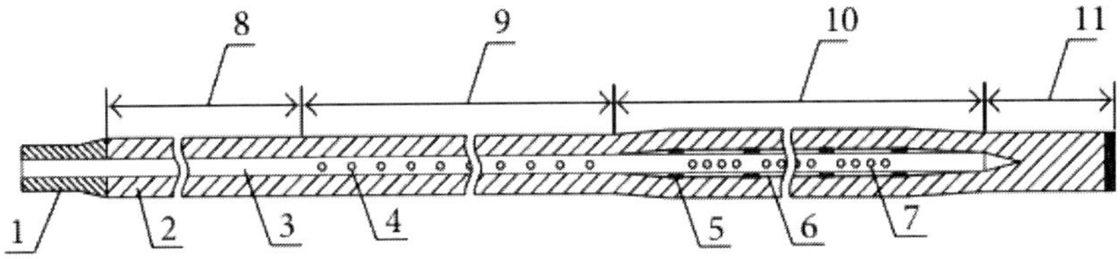


(1) Anchor head
(2) Steel strand
(3) Grouting core tube
(4, 7) Flow hole of slurry
(5) Metal sleeve
(6) Void
(8) Shallow closed area
(9) High pressure grouting section
(10) Low pressure grouting section
(11) Anchoring section

**Figure 11:** Structure of grouting cable.

Compared to the old support design, the new support design can bond the support with the surrounding rock of the roadway and fully utilize the self-supporting capability of the surrounding rock.

# 6. ON-SITE EXPERIMENTS AND MONITORING

Field industrial experiments and monitoring were performed in the −650 main roadway in the south wing of the Yangcheng Coal Mine to test the effectiveness of the new support design and compare it with the old support design. One 100 m segment of roadway was chosen as a trial of the new support design, and the other roadways were supported with the old support design shown in Figure 3. The deformation of the roadway was measured by extensometers with 0.2 mm accuracy. A total of five roadway cross sections were monitored at an interval of 30 m between two sections. Monitoring sections 1, 2, and 3 were in the roadway with the new support design, and monitoring sections 4 and 5 were in the roadway with the old design. The locations of the monitoring sections are shown in Figure 5.

The measuring method of each cross section is shown in Figure 12. In Figure 12, B and E are two vertical data marks fixed in the roof and floor of the roadway, respectively, and C and D are two horizontal data marks fixed in the ribs. CD is the horizontal reference line, and BE is the vertical measuring line. A is the node of CD and BE. A decreasing length of CD means shrinkage of the two ribs, a decreasing length of BA means settlement of the roof and a decreasing length of AE means heaving of the floor.

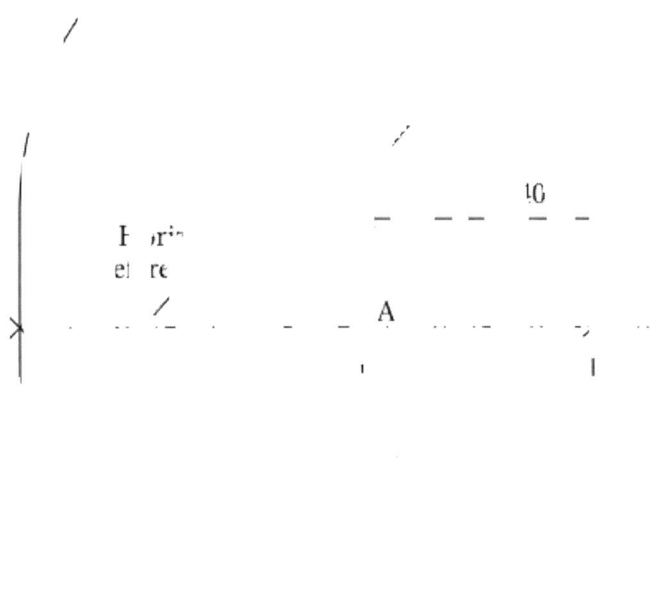


**Figure 12:** Monitoring sites for the deformation of the roadway.

Monitoring of all five roadway cross sections started as soon as possible after all of the supports were installed. Monitoring for sections 1 to 3 lasted for 90 days. Deformation and the deformation rate were both very large in the roadway with the old support design; therefore, to ensure safety of the workers, sections 4 and 5 were monitored for only 70 days.

The monitoring results are shown in Figures 13, 14, 15, and 16. The monitoring results of sections 1 to 3 showed that there was an obvious increase in the deformation during the first 12 days after supporting the roadway. However, after approximately 12 days, the deformation rates started to decrease, and the deformations appeared to be stable after approximately 30 to 60 days. For the comparable sections 4 and 5 with the old support design, the deformation continued to increase over the whole monitoring period, and no obvious decline in the deformation rates was observed during the monitoring period.

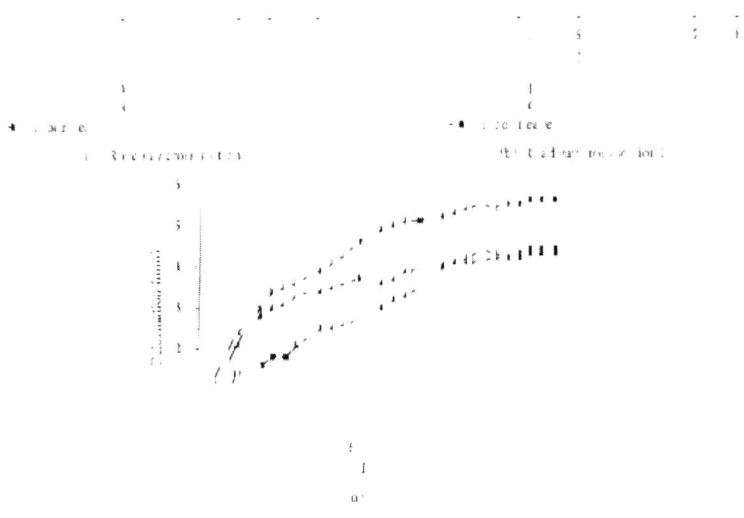

**Figure 13:** The cumulative deformation of the roadway with the new support design.

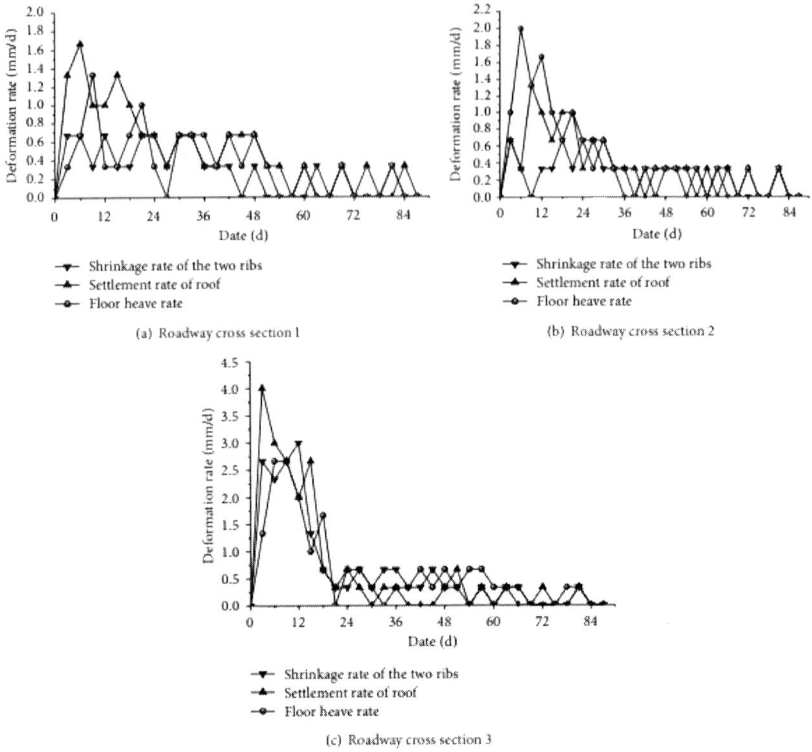


**Figure 14:** The rate of deformation of the roadway with the new support design.

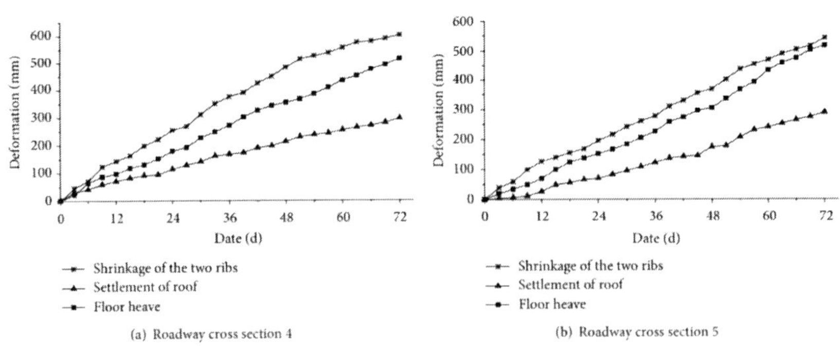


**Figure 15:** The cumulative deformation of the roadway with the old support design.

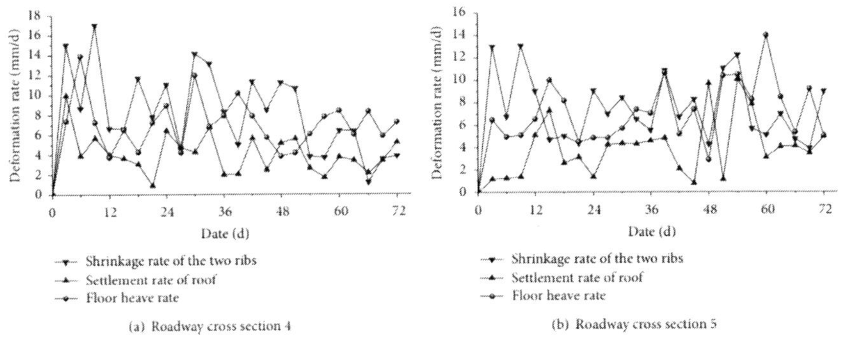


**Figure 16:** The rate of deformation of the roadway with the old support design.

After monitoring, the maximum monitored ribs convergence, roof settlement, and floor heave in the roadway with the new support design were 56 mm (section 3), 44 mm (section 1), and 44 mm (section 3), respectively. In sections 4 and 5 with the old support design, these results were 601 mm (section 4), 295.6 mm (section 4), and 517.2 mm (section 5), respectively. It is clear that the roadway deformation with the new support design is much smaller than that with the old design. A similar conclusion is drawn by comparing the deformation rates of the ribs convergence, roof settlement, and floor heave in the roadway with the new support design and the old design. The highest deformation rates of the two ribs, roof, and floor of the roadway with the new support design were 3 mm/d, 4 mm/d, and 2.7 mm/d, respectively, compared to 17 mm/d, 10 mm/d, and 14 mm/d in the roadway with the old design.

In general, compared with the old support design, the new support design can reduce the roadway deformation by at least 85–90%. During the monitoring period, the roadway roof, ribs, and floor with the new support design were intact and no obvious failure was visible. In contrast, the roadway with the old design experienced large deformations and severe failure. This indicates that the new support design can effectively control the deformation of the roadway and maintain stability of the roadway. The effectiveness of the new support design can be visually observed in Figure 17.

**Figure 17:** The roadway with new supporting design.

## 7. CONCLUSIONS

A case study of the failure mechanisms and support technology for deep composite soft rock roadways was performed on the Yangcheng Coal Mine in Shandong Province, China. The buried depth of the main roadway was approximately 700 m, and the surrounding rock of the roadway was weak and broken. Bolts combined with cables, metal mesh, and even U-shaped steel sheds were used as permanent supports to control the surrounding rock. However, the main roadway still suffered large deformations and seriously unstable failures.

Both a field survey and an experimental study showed that the surrounding rock of the roadway was an HJS complex soft rock characterized by poor rock quality, wide development of joint fissures, and an unstable creep property. The broken-rock zone of the roadway surrounding rock was large and the maximum depth of the fissures exceeded 8 m. The tectonic stress field test showed that the major horizontal stress was approximately 1.59 times greater than the vertical stress. The angle between the axis of the main roadway and the major horizontal stress was 87.7°. The weak surrounding rock and high tectonic stress were the main internal causes of roadway instabilities, and the unreasonable support design was the external cause.

Based on the results of experiments, field tests, and extensive analysis, a new support design was proposed, which included a reasonable support arrangement, long cables, grouting reinforcement, high pretensioning of bolts and cables, and overbreak and bolt-grouting backfill for the floor support. Furthermore, a new grouting cable was invented specifically to control the roadway deformation and failure.

An on-site industrial experiment of the new support design was performed in a roadway section approximately 100 m long. The monitoring results confirmed that the new support design was much more effective than the old design. The new support design reduced the deformations of the roadway by at least 85–90% compared to that of the previous roadway. This case study indicated that a deep composite soft rock roadway similar to the −650 main roadway in the south wing of the Yangcheng Coal Mine could be steadily controlled with an elaborate field investigation, scientific analysis, and reasonable support design. The success of this case provides significant guidance for similar roadway construction engineering projects, which will be welcomed by other coal mines.

## ACKNOWLEDGMENTS

This work was supported by the National Basic Research Program of China (2014CB046905), the National Natural Science Foundation of

China (51274191, 51404245, 51374140, and 51204159), and Doctoral Fund of Ministry of Education of China (20130095110018). The authors thank the anonymous referees for their careful reading of this paper and valuable suggestions.

## REFERENCES

1. B. Shen, "Coal mine roadway stability in soft rock: a case study," Rock Mechanics and Rock Engineering, vol. 47, no. 6, pp. 2225–2238, 2013.

2. M. C. He, H. H. Jing, and X. M. Sun, Soft Rock Engineering Mechanics, Science Press, Beijing, China, 2002.

3. Y. L. Lu, L. G. Wang, F. Yang, Y. Li, and H. Chen, "Post-peak strain softening mechanical properties of weak rock," Chinese Journal of Rock Mechanics and Engineering, vol. 29, no. 3, pp. 640–648, 2010.

4. M. C. He, G. F. Li, J. Wang, and J. Cai, "Study on supporting design for large area serious roof caving of deep soft rock roadway in Xing'an coal mine," Chinese Journal of Rock Mechanics and Engineering, vol. 26, no. 5, pp. 959–964, 2007.

5. Y. P. Wu, C. Wang, M. P. Li, Y. F. Zeng, and C. H. Huang, "Mechanism of roof-floor shearing deformation and breakage in coalmine of soft rock roadway," Journal of Xi'an University of Science and Technology, vol. 27, no. 4, pp. 539–543, 2007.

6. C. Li, Z. L. Wang, and T. Liu, "Principle and practice of coupling support of double yielding shell of soft rock roadway under high stress," International Journal of Mining Science and Technology, vol. 24, no. 4, pp. 513–518, 2014.

7. B. Meng, H.-W. Jing, K.-F. Chen, H.-J. Su, S.-Q. Yang, and Y.-H. Li, "Shear and slip failure mechanism and control of tunnels with weak

surrounding rock," Chinese Journal of Geotechnical Engineering, vol. 34, no. 12, pp. 2255–2262, 2012.

8. H. Salari-Rad, M. Mohitazar, and M. R. Dizadji, "Distinct element simulation of ultimate bearing capacity in jointed rock foundations," Arabian Journal of Geosciences, vol. 6, no. 11, pp. 4427–4434, 2013.

9. R. Seedsman, "The stress and failure paths followed by coal mine roofs during longwall extraction and implications to tailgate support," in Proceedings of the 20th International Conference on Ground Control in Mining, pp. 42–49, Morgantown, WVa, USA, 2001.

10. L. Yuan, J.-H. Xue, Q.-S. Liu, and B. Liu, "Surrounding rock stability control theory and support technique in deep rock roadway for coal mine," Journal of the China Coal Society, vol. 36, no. 4, pp. 535–543, 2011.

11. L. Wu, C. Cui, N. Geng, and J. Wang, "Remote sensing rock mechanics (RSRM) and associated experimental studies," International Journal of Rock Mechanics and Mining Sciences, vol. 37, no. 6, pp. 879–888, 2000.

12. W. P. Huang, Y. F. Gao, and J. Wang, "Deep rock tunnel's long large deformation mechanism and control technology under disturbance effects," Journal of China Coal Society, vol. 39, no. 5, pp. 822–828, 2014.

13. H. P. Kang, J. Lin, and Y. Z. Wu, "Development of high pretensioned and intensive supporting system and its application in coal mine roadways," Procedia Earth and Planetary Science, vol. 1, pp. 479–485, 2009.

14. Y.-Y. Jiao, L. Song, X.-Z. Wang, and A. Coffi Adoko, "Improvement of the U-shaped steel sets for supporting the roadways in loose thick

coal seam," International Journal of Rock Mechanics and Mining Sciences, vol. 60, pp. 19–25, 2013.

15. Y. L. Lu, L. G. Wang, and B. Zhang, "An experimental study of a yielding support for roadways constructed in deep broken soft rock under high stress," Mining Science and Technology, vol. 21, no. 6, pp. 839–844, 2011.

16. C. Wang, Y. Wang, and S. Lu, "Deformational behaviour of roadways in soft rocks in underground coal mines and principles for stability control," International Journal of Rock Mechanics and Mining Sciences, vol. 37, no. 6, pp. 937–946, 2000.

17. L.-G. Wang, M.-Y. Li, and X.-Z. Wang, "Study of mechanisms and technology for bolting and grouting in special soft rock roadways under high stress," Chinese Journal of Rock Mechanics and Engineering, vol. 24, no. 16, pp. 2889–2893, 2005.

18. L.-G. Wang, H.-L. Li, and J. Zhang, "Numerical simulation of creep characteristics of soft roadway with bolt-grouting support," Journal of Central South University of Technology (English Edition), vol. 15, no. 1, supplement, pp. 391–396, 2008.

19. H. Lisa, G. Gunnar, F. Åsa, and N. Tommy, "A statistical grouting decision method based on water pressure tests for the tunnel construction stage—a case study," Tunnelling and Underground Space Technology, vol. 33, pp. 54–62, 2013.

20. S. Nadimi and K. Shahriar, "Experimental creep tests and prediction of long-term creep behavior of grouting material," Arabian Journal of Geosciences, vol. 7, no. 8, pp. 3251–3257, 2014.

21. W. P. Huang, Y. F. Gao, and J. Wang, "Deep rock tunnel's long large deformation mechanism and control technology under disturbance effects," Journal of China Coal Society, vol. 39, no. 5, pp. 822–828, 2014.

22. X. M. Sun, D. Wang, C. Wang, X. Liu, B. Zhang, and Z. Liu, "Tensile properties and application of constant resistance and large deformation bolts," Chinese Journal of Rock Mechanics and Engineering, vol. 33, no. 9, pp. 1765–1771, 2014.

23. S. R. Torabi, H. Shirazi, H. Hajali, and M. Monjezi, "Study of the influence of geotechnical parameters on the TBM performance in Tehran-Shomal highway project using ANN and SPSS," Arabian Journal of Geosciences, vol. 6, no. 4, pp. 1215–1227, 2013.

24. W. J. Wang, G. Peng, and J. Huang, "Research on high-strength coupling support technology of high stress extremely soft rock roadway," Journal of the China Coal Society, vol. 36, no. 2, pp. 223–228, 2011.

25. X. Huang, Q.-S. Liu, and Z. Qiao, "Research on large deformation mechanism and control method of deep soft roadway in Zhuji coal mine," Rock and Soil Mechanics, vol. 33, no. 3, pp. 828–834, 2012.

26. N. M. Esfahani and O. Asghari, "Fault detection in 3D by sequential Gaussian simulation of Rock Quality Designation (RQD)," Arabian Journal of Geosciences, vol. 6, no. 10, pp. 3737–3747, 2013.

27. M. R. Shen and J. F. Chen, Rock Mechanics, Tongji University Press, Shanghai, China, 2006.

28. Q. S. Li, F. L. Bo, X. Y. Ding, et al., "Deformation characteristics of soft rock roadway in deep mines and supporting control technology," Journal of Shandong University of Science and Technology (Natural Science), vol. 29, no. 4, pp. 8–14, 2010.

29. S. C. Li, H. P. Wang, Q. H. Qian et al., "In-situ monitoring research on zonal disintegration of surrounding rock mass in deep mine roadways," Chinese Journal of Rock Mechanics and Engineering, vol. 27, no. 8, pp. 1545–1553, 2008.

30. Y.-L. Lu, L.-G. Wang, B. Zhang, and Y.-J. Li, "Optimization of bolt-grouting time for soft rock roadway," Rock and Soil Mechanics, vol. 33, no. 5, pp. 1395–1401, 2012.

# CHAPTER 10

# Estimating Forest Biomass Dynamics by Integrating Multi-Temporal Landsat Satellite Images with Ground and Airborne LiDAR Data in the Coal Valley Mine, Alberta, Canada

**Nasem Badreldin and Arturo Sanchez-Azofeifa**

*Earth and Atmospheric Sciences Department, University of Alberta, Edmonton, AB T6G 2E3, Canada*

## ABSTRACT

Assessing biomass dynamics is highly critical for monitoring ecosystem balance and its response to climate change and anthropogenic activities. In this study, we introduced a direct link between Landsat vegetation spectral indices and ground/airborne LiDAR data; this integration was established to estimate the biomass dynamics over various years using multi-temporal Landsat satellite images. Our case study is located in an area highly affected by coal mining activity. The normalized difference vegetation index (NDVI), enhanced vegetation index (EVI and EVI2), chlorophyll vegetation index (CVI), and tasseled cap transformations were used as vegetation spectral indices to estimate canopy height. In turn, canopy height was used to predict a coniferous forest's biomass using Jenkins allometric and Lambert and Ung allometric equations. The biophysical properties of 700 individual trees at eight different scan stations in the study area were obtained using high-resolution ground LiDAR. Nine models ($H_i$) were established to discover the best relationship between the canopy height model (CHM) from the airborne LiDAR and the vegetation spectral indices (VSIs) from Landsat images for the year 2005, and $HB_9$ (Jenkins allometric equation) and $HY_9$ (Lambert and Ung allometric equation) proved to be the best models ($r^2 = 0.78$; root mean square error (RMSE) = 44 Mg/H, $r^2 = 0.67$; RMSE = 58.01 Mg/H, respectively; $p < 0.001$) for estimating the canopy height and the biomass. This model accurately captured the most affected areas (deforested) and the reclaimed areas (forested) in the study area. Five years were chosen for studying the biomass change: 1988, 1990, 2001, 2005, and 2011. Additionally, four pixel-based image comparisons were analyzed (*i.e.*, 1988–1990, 1990–2005, 2005–2009, and 2009–2011), and Mann-Kendall statistics for the subsets of years were obtained. The detected change showed that, in general, the environment in the study area was recovering and regaining its initial biomass after the dramatic decrease that occurred in 2005 as a result of intensive mining activities and disturbance.

## KEYWORDS

biomass; Landsat; LiDAR; spectral vegetation indices; land cover change

## 1. INTRODUCTION

Monitoring forest biomass dynamics is known as one of the important factors for environmental modeling [1], and obtaining information on forest biomass dynamics has become a necessity for forest management [2]. Additionally, improving the monitoring techniques to assess forest biomass change is required to understand anthropogenic effects on ecosystem balance [3].

In the last decade, the use of passive and active remote sensing sensors to monitor forest biomass has received attention due to their ability to model the carbon cycle, deforestation and forest degradation [4,5,6]. The recent advances in ground and airborne light detection and ranging (LiDAR) remote sensing enable more accurate data acquisition for the vertical and horizontal vegetation structure and better estimation of forest biomass [7,8]. Many investigations have used LiDAR data to estimate biomass by considering the statistical relationship between field measurements and airborne LiDAR [7,9,10]. Data from high-resolution ground LiDAR will maximize accuracy and minimize uncertainty in the findings [11,12,13]. The main purpose of using LiDAR data in our investigation is to measure the biophysical parameters of trees, such as the diameter at breast height (DBH), tree height and canopy width. Multi-spectral optical remote sensing, such as Landsat Thematic Mapper (TM) satellite images, is a valuable data source in ecological research for estimating multi-temporal biomass dynamics and has been widely used for monitoring vegetation cover [14,15,16,17]. Also, Pflugmacher *et al.* [3] discussed the limitations of estimating biomass using multi-spectral sensors, namely, the misrepresentation of the actual spatial distribution of the biomass. Additionally, the difficulty of evaluating the sampling errors because of the spatial resolution heterogeneity between remote sensing data, field surveys, and biomass maps has been discussed by [18,19]. Many scientists succeeded in enhancing the relationship between Landsat spectral indices and forest biomass by incorporating the spectral trajectory of forest disturbance and regrowth, which decreased the root mean square error (RMSE) from 57% to 41% [20].

The allometric equations are useful for estimating biomass in North America, and were collected from different research papers for various tree species [21,22,23]. Equations that are suitable for a large number of species have been developed [24,25] to correct the allometric equations previously published, such as in [26,27,28,29]. Additionally, many scientists have adopted various allometric equations in their case studies because they can be used to address the spatial variability [28,30,31].

Our case study involves a mining area in the western central area of the province of Alberta in Canada, where huge areas were affected by deforestation as a result of coal mining activities. Such anthropological activity introduce abiotic chemical and physical disturbance such as pH (soil acidity and alkalinity), soil salinity, soil compaction, and low hydraulic conductivity, and these factors harm the soil fertility and the ecosystem balance and cause degradations over time [32,33,34].

The main objective of this investigation is to obtain the multi-spatiotemporal biomass distribution from optical remote sensing data for a coniferous forest on a coal mining site. To achieve this goal, we first addressed several research challenges:

(a) Using high-resolution ground LiDAR to find the relationship between different biophysical parameters, including tree height, DBH and canopy width,

(b) Applying a ground-LiDAR-based regression model to two allometric equations to estimate the biomass values from the CHM of the airborne LiDAR,

(c) Finding the best spectral-based regression model to predict the coniferous forest biomass from multi-temporal Landsat data, and

(d) Assessing the change in the biomass in 1988, 1990, 2001, 2005, and 2011.

## 2. CASE STUDY DESCRIPTION

The case study was conducted in the Coal Valley Mine area, which is located approximately 80 km south of Edson in west central Alberta, Canada, as shown in Figure 1. This study site is approximately 7000 Ha. The study area has been used for coal mining since 1978. The natural vegetation in the study area is dominated by coniferous forests composed of *Pinus contorta* Loudon (lodgepole pine) with ericaceous shrubs and feather mosses. *Populus tremuloides* Michx. (Aspen) and mixed Pinus and Populus stands with herbaceous and deciduous shrub understories are present on the south aspects and the crests of knolls [35,36]. Long-term continuous climatic data are not available for our particular study site, but the average climatic data of the precipitation, temperature and snowfall were obtained from the nearest NOAA station (Id: CN71881030622440) in Edson (53°35'N, 116°28'W) for the period of 1961–2010 [37]. The average annual total precipitation is approximately 570 mm, and 305 mm occurs June–August. The coldest period is December-January, with a mean temperature of −18 °C, and the warmest period is July–August, with a mean temperature of 20 °C. The highest snowfall is 35 cm in January, and the lowest snowfall is approximately 6–8 cm in May and September, according to the preliminary field surveys that were conducted on 13 June 2013 and 28 April, 20 May, 11 July, 5 August 2014. Reclamation activities included the placement of cover soil, fertilization, and seeding with a mixture of legume and graminoid species; in the future, shrubs and tree seedlings may be planted to reforest the former mining sites.

## 3. MATERIALS AND METHODS

This investigation is based on various remote sensing data that were acquired and integrated via advanced analytical techniques, as shown in Figure 2. The spatial analyses were conducted using several datasets (*i.e.*, multi-temporal Landsat satellite images, high-resolution ground LiDAR, and airborne LiDAR). The integration system analyses were conducted after the pre-processing data quality check to remove the noise and unify the data geo-references. Software packages, such as ArcGIS 10.1, Cyclone 8.11, and ENVI 5.0, were used to achieve the objectives.

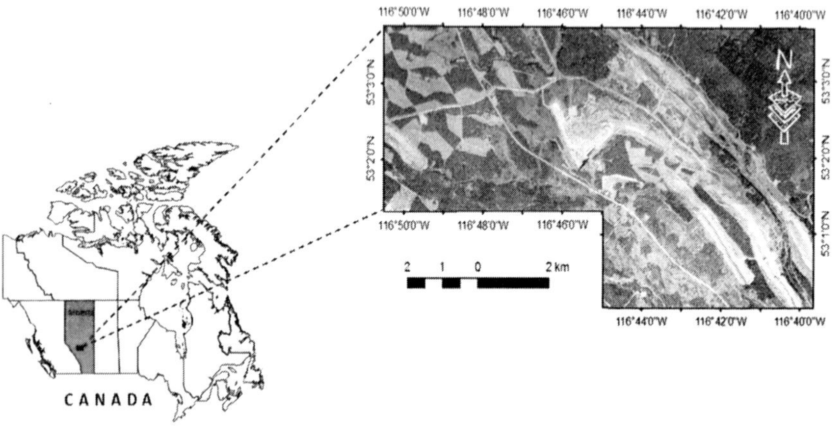


**Figure 1.** The geographical location of the case study, aerial imagery with 0.5 m spatial resolution.

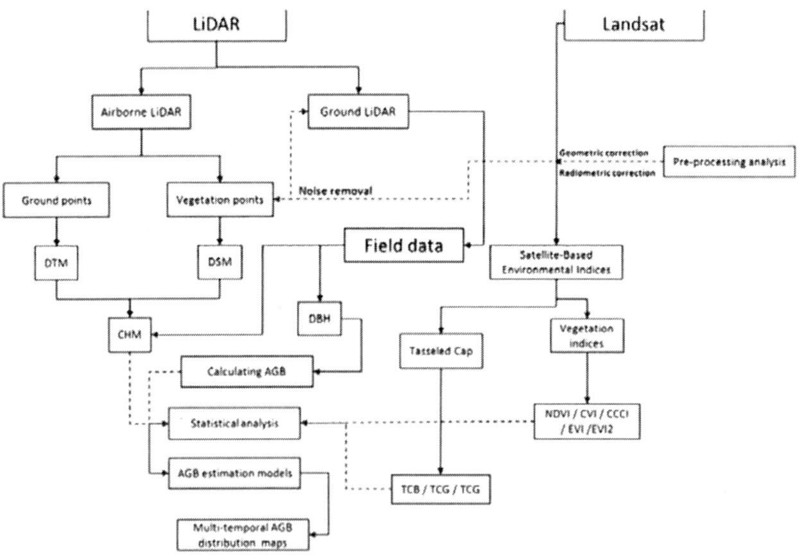


**Figure 2.** Flow chart describing the integration and the analyses steps of Landsat and LiDAR data.

## 3.1. Airborne LiDAR and Canopy Height Extraction

The discrete airborne LiDAR data for the case study were acquired on 22 November 2005 and 26 March 2006. The horizontal root-mean-square error (RMS) is 45 cm, and the vertical RMS is 30 cm. The point density of the acquired LiDAR data for all tiles is greater than 1.3 points per square meter. Environmental System Research Institute (ESRI) ArcGIS 10.1 was used to process the LAS v1.0 point cloud, which contained all return point clouds. Digital elevation models (DEM) and digital surface models (DSM) were produced after removing the noise from the data acquisition. A canopy height model (CHM) with one meter spatial resolution was calculated by subtracting the DEM from DSM; the CHM represents a 3D surface of the vegetation cover.

## 3.2. Ground LiDAR and Field Measurements

The forest biophysical parameters were collected on 20 May and 11 July 2014, using Leica ScanStation C10. An exploratory field survey was conducted on 28 April 2014, to locate the scan stations that represent the overall ecosystem of the case study. Eight scan stations were placed at three locations, where the range of the trees metrics was 300 m and the scan rate was 50,000 points/s. The scan stations are fully operational in conditions ranging from bright sunlight to complete darkness, and the accuracy of single tree's (object) measurements was 4–6 mm. Cyclone 8.11 software was used to calculate the following characteristics of individual trees: diameter at breast height (DBH), height (H), and canopy width. One of the major advantages of using ground LiDAR is that the obtained data do not require extra field surveys for validation [38,39].

## 3.3. Landsat Data

Landsat 5 Thematic Mapper TM satellite images in cloud-free conditions were obtained from the US Geological Survey for spring and summer days in five years, *i.e.*, 2 September 1988, 15 September 1990, 28 May 2005, 27 August 2009, and 9 September 2011. The Landsat data were re-projected into a Universal Transverse Mercator (UTM), zone 11 north NAD 1983 horizontal coordinate system. ENVI version 5.0 was used to apply the radiometric correction analysis to all multi-temporal Landsat data using the Landsat calibration utility for atmospheric correction. The calibrated images provide better reflectance for minimizing the climate effects that could potentially lead to miscalculations or additional noise. The small temporal heterogeneity (less than three months) between the Landsat dataset will not affect the change detection analysis because of the slow rate of growth in coniferous species [40,41].

## 3.4. Calculating the Spectral Indices

The normalized difference vegetation index (NDVI), enhanced vegetation index (EVI and EVI2), chlorophyll vegetation index (CVI), and tasseled cap transformation were considered as vegetation spectral indices (VSIs). These indices were estimated from the multi-temporal Landsat satellite data (Table 1), which were integrated with the ground LiDAR measurements to obtain the biophysical parameters of the studied forest; the parameters were used to estimate the biomass. The VSIs were statistically related to the biomass status and spatial distribution and were resistant to atmospheric influences and the forest structure [16,17,42,43,44].

Tasseled cap transformations are considered to be robust indicators of forest dynamics [50]. A linear combination of sensor-specific weightings for each band (Table 2) was applied to the multi-temporal Landsat 5 satellite images; these transformations are the three orthogonal indices called brightness, greenness and wetness (TCB, TCG and TCW, respectively) [20,51].

**Table 1.** The derived VSIs equations from Landsat 5 TM, ρNIR is the near infrared band, ρR is the red band, ρB is the blue band, and ρG is the green band. *G*, $C_1$, $C_2$, and *L* are 2.5, 6, 7.5, and 1 respectively.

| Spectral Index | Formula | Reference |
|---|---|---|
| ***NDVI*** | $\frac{\rho_{NIR} - \rho_R}{\rho_{NIR} + \rho_R}$ | [45] |
| ***EVI*** | $G\frac{\rho_{NIR} - \rho_R}{\rho_{NIR} + C_1\rho_R - C_2\rho_B + L}$ | [46,47] |
| ***EVI2*** | $2.5\frac{\rho_{NIR} - \rho_R}{\rho_{NIR} + 2.4\rho_R + L}$ | [15] |
| ***CVI*** | $\frac{\rho_{NIR} \times \rho_R}{\rho_G^2}$ | [48,49] |

**Table 2.** Landsat 5 TM band-specific weights for TCB,TCG and TCW.

| Tasseled Cap Feature | Weighting for Tasseled Cap Transformation Using Landsat 5 TM Dataset | | | | | |
|---|---|---|---|---|---|---|
| | B1 | B2 | B3 | B4 | B5 | B7 |
| Brightness | 0.3037 | 0.2793 | 0.4343 | 0.5585 | 0.5082 | 0.1863 |
| Greenness | −0.2848 | −0.2435 | −0.5436 | 0.7243 | 0.0840 | −0.1800 |
| Wetness | 0.1509 | 0.1793 | 0.3299 | 0.3406 | −0.7112 | −0.4572 |

## 3.5. Allometric Equations and Ground LiDAR Metrics from Individual Trees and Point Clouds

The Jenkins allometric equations model are suitable for predicting the biomass based on DBH for different coniferous species [26,27,28], Jenkins *et al.* [22] estimated the biomass as follow:

$$B_i = exp(\beta_0 + \beta_1\, ln(DBH)) \quad (1)$$

where Bi is the biomass (kg), DBH is the diameter in meters at breast height (1.3 m), ln is the natural log base, and β0, β1 are the coefficients.

Also, Lambert *et al.* [52] and Ung *et al.* [53] provided allometric equations to estimate biomass were derived from thousands of samples (trees) across Canada [54]. The allometric equation form is shown as follows:

$$Y_i = \beta_1 DBH^{\beta_2} H^{\beta_3} \quad (2)$$

where Yi is the biomass (kg), β1, β2, β3 are the coefficients.

Ni-Meister *et al.* [8] Found that LiDAR metrics can be used as predictors for biomass. In this study, we used a new high-resolution ground LiDAR to obtain the relationship between the DBH and the tree height (H) in tree-level metric bases ($r^2 = 0.74$; $p < 0.001$). Approximately 700 individual trees in the coniferous forest were obtained in the study site (Figure 3), and the empirical relationships from the ground LiDAR field measurements were developed as follows:

$$DBH = 0.098 \times H - 0.055 \quad (3)$$

where DBH is the diameter in meters at breast height (at 1.3 m) and H is the tree height in meters.

**Figure 3.** A 3D visualization of the data obtained from the high resolution ground LiDAR.

## 3.6. Statistical and Temporal Trend Analysis

The stepwise multiple linear regression method was used to select the most significant variables; multiple linear regressions were conducted for airborne LiDAR and derived VSI from Landsat 5 data for 2005. We determined the most suitable model, which is based on the linear regression scatter plot with the ground LiDAR results that identified the biophysical metrics for the study. This model was used to predict H for the years of 1988, 1990, 2009, and 2011 and to map the biomass in the study years. The conventional multivariate regression model can be expressed as follows:

$$y = x_0 + a_1 x_1 + \cdots + a_i x_i \quad (4)$$

where y is CHM in meters and is considered to be the dependent parameter to be predicted, x0 is the intercept, a1x1 are the regression coefficients and the values of the independent variables, and i is the number of independent variables.

To determine the model that best predicted the biomass spatial distribution, 409 points were used to validate the prediction models; we used the root mean square error (RMSE), which indicates how far the predicted values deviated from the measured data (ground LiDAR) [55].

$$RMSE = \sqrt{\frac{1}{N}\sum_{i=1}^{N}(y_0 - y_i)^2} \quad (5)$$

where y0 is the predicted values, yi is the measured values from the ground LiDAR, and N is the number of the observations and grid points.

The Akaike's Information Criterion (AIC) can be defined as the objective way to decide between models that provide best fits, and recently used as an assessment tool to select the best model that predict/or estimate phenomena [56].

$$AIC=-2\ln L+2k \quad (6)$$

where lnL is the maximized log-likelihood of the model and k is the number of parameters estimated [57].

$$\Delta AIC_i = AIC_i - AIC_{min} \quad (7)$$

Also, AIC differences can be used to identify the best model that has$\Delta AICi \equiv 0$, where AICi is the Akaike of model i and AICmin is the lowest Akaike among the tested models i.

A pixel-based image comparison was used to estimate the amount of change over the study years. Four temporal subsets were analyzed: 1988–1990, 1990–2005, 2005–2009, and 2009–2011. The comparisons were based on the difference between the same spatial pixel positions for each pair of multi-year periods. The difference between each pair provides useful data on the most degraded or reclaimed sites in the study area [58,59]. A temporal trend analysis was provided by [17]. The Mann-Kendall trend test analysis is a nonparametric test for linear regression with zero slopes. Each pixel location was calculated in a time series; the S statistics increase when the value of the later data is higher than the earlier data, and the cumulative score of all pairs is used to calculate S [60]. Gilbert [61] proposed the following procedures to calculate the Mann-Kendall trend statistics:

(1) The biomass data should be listed in temporal order:

$$x_1, x_2, \dots, x_n$$

where xi is the biomass estimated maps at time i

(2)The sign (xj–xk) is the indicator function, where the value of all n(n–1)/2 must be determined for all differences of xj–xk where j>k.

$$\text{sign}(x_j - x_k) = \begin{cases} +1 \; if \; x_j - x_k > 0 \\ 0 \; if \; x_j - x_k = 0 \\ -1 \; if \; x_j - x_k < 0 \end{cases} \tag{8}$$

(3)Computing the Mann-Kendall Statistics as follow:

$$S = \sum_{k=1}^{n-1} \sum_{j=k+1}^{n} \text{sign}\,(x_j - x_k) \tag{9}$$

# 4. RESULTS

## 4.1. Regression Models and Biomass Estimation

Stepwise multiple regression analysis was used to estimate the canopy height from the Landsat spectral indices (NDVI, EVI2, TCW, TCG, and TCB), which are independent variables. Nine models were established to identify the best relationship between the airborne LiDAR and Landsat data for 2005 from 4771 points that were extracted from all variables at the same location. $H_8$ and $H_9$ were the best models for estimating the canopy height ($r^2$ = 0.73, and 0.74; $p <$ 0.001, respectively. ΔAICi=0 and 176.4, respectively), as shown in Table 3.

The final equations that were used to predict the biomass directly from canopy height are shown as follow:

$$B_i = exp(0.1617 + 2.142 \times Ln[0.098 \times H_i - 0.055]) \tag{10}$$

$$Y_i = 22.82 \times (0.098 \times H_i - 0.055)^{0.05612} H_i^{\,0.8931} \tag{11}$$

where Bi is the biomass in $Mg{\cdot}H^{-1}$ based on the Jenkins allometric equation, Yi is the biomass in $Mg{\cdot}H^{-1}$based on the Lambert and Ung allometric equation, and Hi is the canopy height in meters.

**Table 3.** A summary of multiple regression analysis results and regression models.

| Variables | Models | $R^2$ | $\Delta AIC_1$ |
|---|---|---|---|
| ***TCB,TCG,TCW*** | $H_1 = 18.67 - 0.13 \times TCB + 0.15 \times TCG - 0.04 \times TCW$ | 0.67 | 1047.81 |
| ***TCB,TCW,NDVI*** | $H_2 = 8.40 + 21.13 \times NDVI - 0.08 \times TCB - 0.04 \times TCW$ | 0.70 | 618.22 |
| ***TCB,TCW,EVI2*** | $H_3 = 8.54 + 8.52 \times EVI2 - 0.08 \times TCB - 0.04 \times TCW$ | 0.70 | 621.35 |
| ***TCW,NDVI*** | $H_4 = -0.75 + 27.36 \times NDVI + 0.02 \times TCW$ | 0.66 | 1265.23 |
| ***TCW,EVI2*** | $H_5 = -0.67 + 10.98 \times EVI2 + 0.02 \times TCW$ | 0.66 | 1287.59 |
| ***TCB,NDVI*** | $H_6 = 9.04 + 16.39 \times NDVI - 0.06 \times TCB$ | 0.69 | 722.51 |
| ***TCB,EVI2*** | $H_7 = 9.17 + 6.58 \times EVI2 - 0.06 \times TCB$ | 0.69 | 727.05 |
| ***TCB,TCG,NDVI*** | $H_8 = -6.12 + 0.01 \times TCB - 0.38 \times TCG + 54 \times NDVI$ | 0.73 | 176.4 |
| ***TCB,TCG,TCW,EVI2*** | $H_9 = -14.79 - 0.63 \times TCG + 28.57 \times EVI2 + 0.09 \times TCB + 0.07 \times TCW$ | 0.74 | 0 |

Figure 4 shows the comparisons between the measured biomass from the high-resolution ground LiDAR and the estimated biomass derived from the Landsat spectral indices models. The models with higher adjusted $r^2$ and lower RMSE are indicative of better biomass estimations from the spectral indices variables; the models ($HB_7$ and $HB_9$) with higher $r^2$ values had lower RMSE values. TCG and TCW increased $r^2$ + 0.1, and the RMSE decreased (−16.87 Mg/H). Also, the Lambert and Ung allometric equation provided the same superior models ($HY_7$ and $HY_9$), see Figure 5.

## 4.2. Estimation Uncertainties

Understanding, identifying, and reducing the sources of uncertainties are critical in improving forest biomass estimation performances. In this investigation, based on the ground LiDAR data survey, a group of statistical models was developed. The models were based on the best relationships between the tree height data, DBH and spectral indices to estimate biomass from the Landsat images. However, our models involve uncertainty for the following reasons. First, a scale difference exists between remotely sensed data and ground LiDAR survey data.

The spatial resolution of Landsat 5 data is 30 m. Considering the high-resolution ground LiDAR scans, the average values of multiple samples were used to reduce the estimation error caused by the scale difference. Second, the satellite image data used in this study originate from Landsat 5 and require an atmospheric correction to obtain the best spectral reflectance for surface greenness and distribution. Third, there are uncertainties concerning the field survey for biomass change detection because simultaneous field data and satellite images are difficult to obtain. To address the temporal variance between the field survey and the remote sensing data acquisition, we focused on the relationships between the biophysical parameters in the study sites rather than the temporal data acquisition; the ultimate goal of our research is to provide an optimal biomass prediction model.

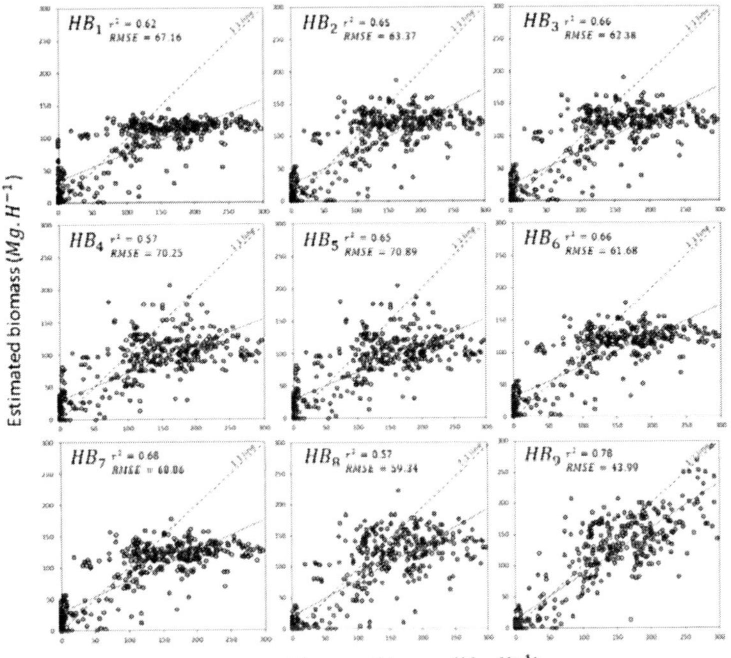


**Figure 4.** The stepwise multiple linear regression results for the measured and estimated biomass analysis of nine different models Hi(see, Table 3), and based on Jenkins allometric equation Bi.

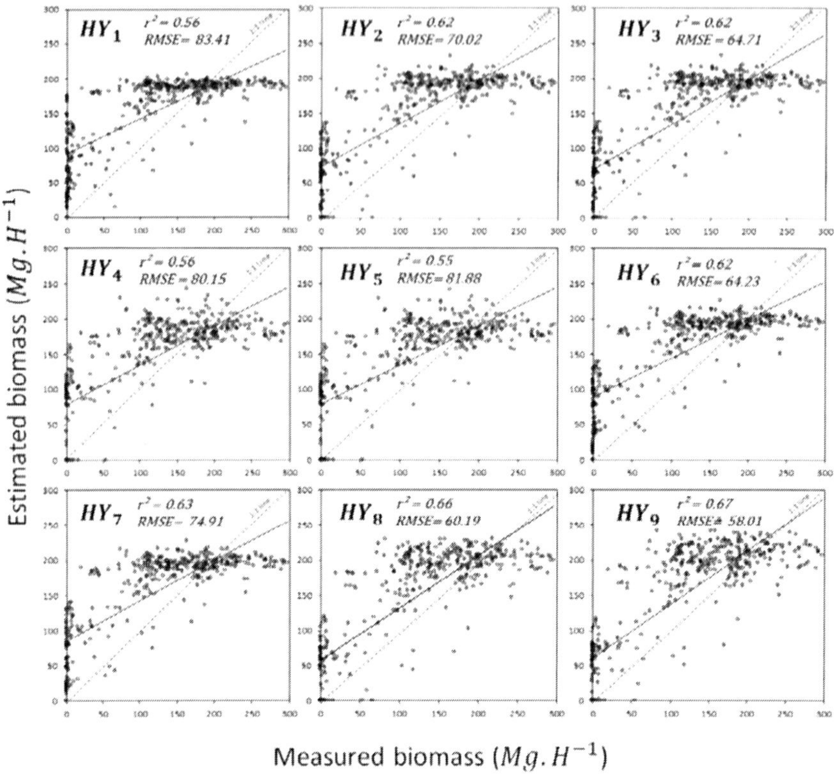


**Figure 5.** The stepwise multiple linear regression results for the measured and estimated biomass analysis of nine different models Hi(see, Table 3), and based on Lambert and Ung allometric equation Yi.

## 4.3. Biomass Spatial Distribution and Trajectory of Change

The estimated biomass results had a wide ranging spatial distribution as a result of mining activities that led to deforestation, as shown in Figure 6. During 1988–1990, the total area with a biomass <100 Mg/H decreased approximately 55.3%, and the biomass in the southeastern area increased approximately 36.8%. The northwestern and eastern areas lost approximately 2.2% of the >600 Mg/H biomass area, which shows the capability of the estimation model that was used to map the biomass spatial distribution. The study site was massively deforested, except for the northeastern side. After 2005, land reclamation occurred. During 2005–2009, the vegetation cover area increased 26.6% for the trees that had biomass >600 Mg/H, 25.4% for biomass of 100–250 Mg/H, and 23% for biomass <100 Mg/H. The biomass of trees <100 Mg/H increased 57.8% as a result of continuous land reclamation and regrowth.

As shown in Figure 7, a trend analysis for 1988–2011 was conducted using the estimated biomass data; the analysis showed a significant decrease in the biomass over this 23-year period. Applying the biomass prediction model to the S statistics provided clear information on how massively the mining activities deforested huge areas for this period. Changing the time window provided a different view of the activities at the same study sites. During 2009–2011, the trend analysis showed an obvious increase in the biomass, which is clear evidence of reclamation activity and regrowth.

# 5. DISCUSSION

## 5.1. Major Findings

The integration of data from optical remote sensing and airborne and ground LiDAR provided the ultimate data source for estimating and mapping the spatial and temporal biomass distribution for coniferous forests. Many scientists have demonstrated that Landsat and LiDAR data can be used to derive highly reliable maps of biomass distributions based on field survey measurements [3,16,50].

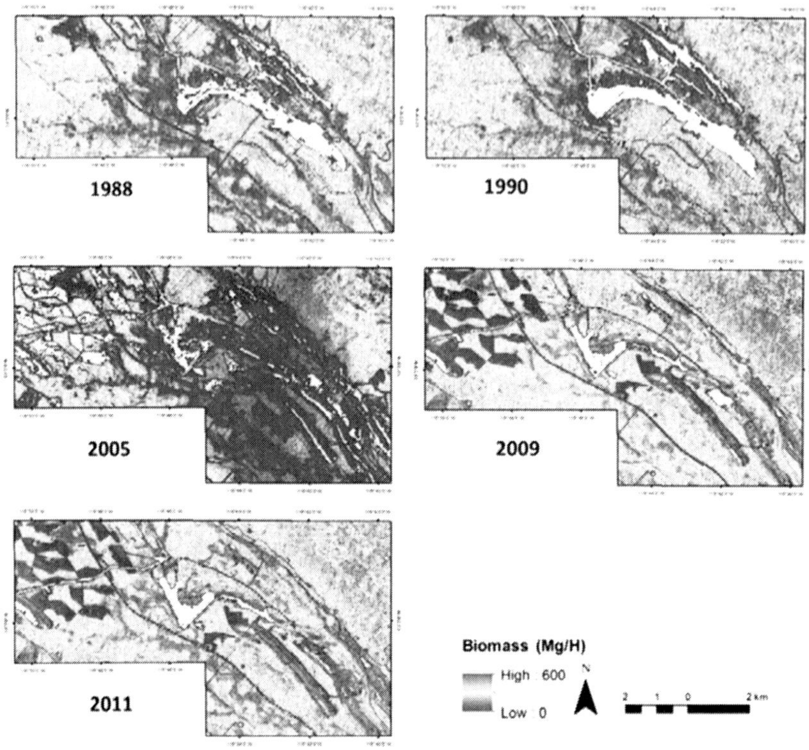


**Figure 6.** The spatial distribution of the estimated biomass over the case study over the years 1988, 1990, 2005, 2009, and 2011; the spatial resolution is 30 m.

In this study, we developed an approach that further integrates airborne LiDAR data with the high-resolution ground LiDAR to specifically determine the relationships between the biophysical characteristics of a forest and spectral indices from Landsat data. The relationships were used to directly estimate the biomass distribution at the study site from the spectral indices that were calculated from multi-temporal Landsat satellite images. Several key elements were considered to estimate and map the biomass temporal dynamics and spatial distribution using Landsat data. First, information and statistical relationships derived from both the airborne and high-resolution ground LiDAR were used to improve the biomass modeling substantially and provide accurate datasets. The advantage of using ground LiDAR for biomass modeling is the capability of accurately measuring all the biophysical details at the study site and of linking the ground LIDAR data to the airborne LiDAR data. This connection revealed information undetectable from airborne LiDAR; thus, this information is essential to provide the best model for estimating biomass directly from optical remote sensing data, such as Landsat satellite data. Second, spectral indices were carefully chosen to provide more information on the vegetation cover, as shown in Table 1 and Table 2. In this research, VSIs were derived from calibrated multi-temporal Landsat images to detect and map the location and magnitude of biomass dynamics. Several studies demonstrated the benefit of using Landsat images for change detection in coniferous forest environments [62]. We provide new information on the potential use and capability of such methods in estimating and mapping biomass dynamics. Third, the modeling approach has a high degree of certainty, based on the direct link between spectral indices and biophysical parameters captured by the airborne and ground LiDAR. TCB, TCG, TCW, EVI2, and NDVI were highly suitable variables for estimating the biomass spatial distribution for our case study, as shown in Table 3. Similarly to [17], a strong, negative relationship was found between brightness reflectance and vegetation coverage ($r = 0.81$); this relationship represents the bare soil spots that result from the mining activities in the study site. The TCW is a highly weighted (positively) spectral variable with CHM ($r = 0.78$); this relationship supported the modeling approach to estimate the biomass according to the canopy heights [13,27]. Fourth, Zhao *et al.*[28] stated that the multiple linear models with height-related metrics estimate the biomass variation better than the profile-related metrics because the biomass is directly related to the height and/or DBH, whereas the canopy-profile-based metrics provide very limited information. Essentially, the Jenkins allometric equations used in this research employs DBH as the input, but it is still possible to use an additional variable (*i.e.*, height), as suggested by [27,28].

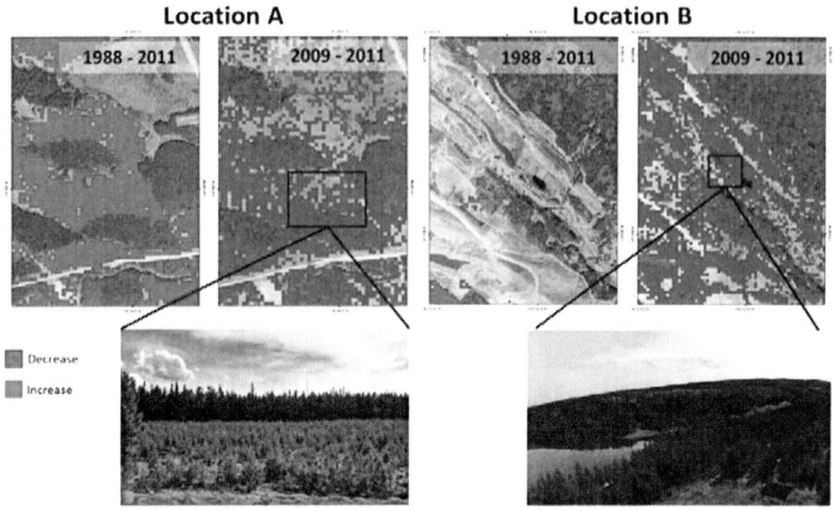


**Figure 7.** A zoomed view of two sites in the case study showing the decrease and increase of the biomass for long-term (1988–2011) and short-term (2009–2011).

Models with higher adjusted $r^2$ and low RMSE likely have a greater ability to predict biomass using LiDAR-derived variables. $HB_9$ and $HY_9$ were the best models regarding biomass estimations using allometric equations. The final regression model $HB_9$ was further used with the multi-temporal Landsat images to estimate the biomass spatial distribution; this model is useful for cases that have intense coniferous forests. Additionally, these results showed the advantages of using Landsat satellite images to map and assess biomass change over time and to detect deforestation and reclamation effects (see Figure 6). Finally, change detection provides more information on the history of the study site in terms of the trajectory of change. Massive anthropogenic activity appeared in 2005 throughout the study site. The northeastern side of the study site, which has a high elevation, showed more vegetation cover stability, which is a morphological characteristic that protects trees from the mining activities. In 2004, mining activities increased to nearly 4 million tons/year, and the mining reached its full extent in the summer of 2006, in which thermal coal production doubled [35]. During 1990–2005, the areas that had a biomass <100 Mg/H decreased to 90.3%, and the areas that had a biomass >600 Mg/H decreased by 7.2%.

## 5.2. Suggestions for Future Research

This research successfully used multi-temporal Landsat data to develop and introduce a new methodology for assessing biomass change in coniferous forests in human-affected areas; nevertheless, some aspects need to be considered in future research. Landsat satellite data provide the most continuous time series for assessing biomass change. However, it is highly preferable to use continuous multi-temporal satellite data in bi-weekly temporal intervals to avoid possible disturbances/or natural hazards (wildfires, floods, *etc.*) that could affect the findings. Integrating different sensors could cause errors due to spatial resolution heterogeneity, and therefore, any integration approaches in environmental research must be taken gradually by developing the integration analysis in a stepwise manner and assessing the accuracy of each step. We believe that integrating ground LiDAR directly with Landsat data will result in more errors rather than greater accuracy. A field survey is essential to verify the findings and to confirm whether the predictions correspond with reality. The use of high-resolution optical satellite sensors (Worldview, Quick Bird, and Spot) is a significant advantage of this type of research, but these data are costly compared with the free Landsat archive.

# 6. CONCLUSIONS

In this study, we estimated the multi-temporal and spatial distribution of biomass for a coniferous forest at a coal mining site to assess the biomass changes as a result of land degradation and reclamation. Various datasets and techniques were used and integrated to provide the best predictors for estimating biomass directly from Landsat data.

Ground LiDAR metrics were essential to understanding and accurately finding the best relationship between different biophysical parameters in our case study. A strong relationship ($r^2 = 0.74$; $p < 0.001$) was found between tree height and DBH; this statistical relationship was based on the metrics of 700 individual trees. The ground-LiDAR-based regression model was used to replace the DBH in Jenkins allometric equation with the tree height, which was obtained from airborne LiDAR. This step helped estimate the biomass from the airborne LiDAR for 2005 directly from the tree height that was extracted from the difference between the DEM and DSM. Nine spectral based models were analyzed with the airborne LiDAR metrics to select the best biomass predictor; $HB_9$ proved to be the best predictor ($r^2 = 0.78$; RMSE = 44 Mg/H; $p < 0.001$). This model was used to estimate the biomass spatial distributions for 1988, 1990, 2001, 2005, and 2011. A major loss of biomass was found in 2005, when the study site lost approximately 50% of its forested area. This major deforestation was the result of extreme excavation. The land morphology in the northeastern side of the site played a vital role in protecting the natural habitats from the excavation activity. The areas with biomass <100 Mg/H were indicators for land reclamation and conservation, both of which increased dramatically during 2005–2011. Predicting biomass from Landsat data requires different resolution levels of data records that can be integrated for understanding the disturbance activity that occurred. With the continuity of the Landsat data, developing new algorithms becomes highly essential for studying historical changes and understanding the connection between past and current environmental states [20].

## ACKNOWLEDGMENTS

We would like to acknowledge Cassidy Rankine, Megan Hill, and Stephanie Autut for their assistance during the field work. Also, we gratefully acknowledge USGS EROS and Shane Patterson for the data support.

## AUTHOR CONTRIBUTIONS

All authors contribute equally in this manuscript.

## REFERENCES

1. He, Q.; Chen, E.; An, R.; Li, Y. Above-Ground biomass and biomass components estimation using LiDAR data in a coniferous forest. *Forests* **2013**, *4*, 984–1002.

2. Fernández-Manso, O.; Fernández-Manso, A.; Quintano, C. Estimation of aboveground biomass in Mediterranean forests by statistical modelling of ASTER fraction images. *Int. J. Appl. Earth Obs. Geoinf.***2014**, *31*, 45–56.

3. Pflugmacher, D.; Cohen, W.B.; Kennedy, R.E.; Yang, Z. Using landsat-derived disturbance and recovery history and LiDAR to map forest biomass dynamics. *Remote Sens. Environ.* **2014**, *151*, 124–137.

4. Deng, S.; Katoh, M.; Guan, Q.; Yin, N.; Li, M. Estimating forest aboveground biomass by combining ALOS PALSAR and WorldView-2 data: A case study at Purple Mountain National Park, Nanjing, China. *Remote Sens.* **2014**, *6*, 7878–7910.

5. Laurin, G.V.; Chen, Q.; Lindsell, J.A.; Coomes, D.A.; del Frate, F.; Guerriero, L.; Pirotti, F.; Valentini, R. Above ground biomass estimation in an African tropical forest with LiDAR and hyperspectral data. *ISPRS J. Photogramm. Remote Sens.* **2014**, *89*, 49–58.

6. Kwak, D.A.; Lee, W.K.; Lee, J.H.; Biging, G.S.; Gong, P. Detection of individual trees and estimation of tree height using LiDAR data. *J. For. Res.* **2007**, *12*, 425–434.

7. Cao, L.; Coops, N.; Hermosilla, T.; Innes, J.; Dai, J.; She, G. Using small-footprint discrete and full-waveform airborne LiDAR metrics to estimate total biomass and biomass components in subtropical forests. *Remote Sens.* **2014**, *6*, 7110–7135.

8. Ni-Meister, W.; Lee, S.; Strahler, A.H.; Woodcock, C.E.; Schaaf, C.; Yao, T.; Ranson, K.J.; Sun, G.; Blair, J.B. Assessing general relationships between aboveground biomass and vegetation structure parameters for improved carbon estimate from lidar remote sensing. *J. Geophys. Res.* **2010**, *115*, G00E11.

9. Edson, C.; Wing, M.G. Airborne Light Detection and Ranging (LiDAR) for individual tree stem location, height, and biomass measurements. *Remote Sens.* **2011**, *3*, 2494–2528.

10. Kim, Y.; Yang, Z.; Cohen, W.B.; Pflugmacher, D.; Lauver, C.L.; Vankat, J.L. Distinguishing between live and dead standing tree biomass on the North Rim of Grand Canyon National Park, USA using small-footprint LiDAR data. *Remote Sens. Environ.* **2009**, *113*, 2499–2510.

11. Dandois, J.P.; Ellis, E.C. Remote sensing of vegetation structure using computer vision. *Remote Sens.***2010**, *2*, 1157–1176.

12. Seielstad, C.; Stonesifer, C.; Rowell, E.; Queen, L. Deriving fuel mass by size class in Douglas-fir (Pseudotsuga menziesii) using terrestrial laser scanning. *Remote Sens.* **2011**, *3*, 1691–1709.

13. Lu, D.; Chen, Q.; Wang, G.; Moran, E.; Batistella, M.; Zhang, M.; Laurin, G.V.; Saah, D. Aboveground forest biomass estimation with landsat and LiDAR data and uncertainty analysis of the estimates. *Int. J. For. Res.* **2012**, *2012*, 1–16.

14. Kelsey, K.; Neff, J. Estimates of aboveground biomass from texture analysis of landsat imagery.*Remote Sens.* **2014**, *6*, 6407–6422.

15. Miura, T.; Yoshioka, H.; Fujiwara, K.; Yamamoto, H. Inter-Comparison of ASTER and MODIS surface reflectance and vegetation index products for synergistic applications to natural resource monitoring.*Sensors* **2008**, *8*, 2480–2499.

16. Chen, X.; Vierling, L.; Rowell, E.; DeFelice, T. Using LiDAR and effective LAI data to evaluate IKONOS and Landsat 7 ETM+ vegetation cover estimates in a ponderosa pine forest. *Remote Sens. Environ.* **2004**, *91*, 14–26.

17. Czerwinski, C.J.; King, D.J.; Mitchell, S.W. Mapping forest growth and decline in a temperate mixed forest using temporal trend analysis of Landsat imagery, 1987–2010. *Remote Sens. Environ.* **2014**, *141*, 188–200.

18. Mitchard, E.T.A; Saatchi, S.S.; Lewis, S.L.; Feldpausch, T.R.; Gerard, F.F.; Woodhouse, I.H.; Meir, P. Comment on A first map of tropical Africa's above-ground biomass derived from satellite imagery.*Environ. Res. Lett.* **2011**, *6*, 49001.

19. Zhang, G.; Ganguly, S.; Nemani, R.R.; White, M.A.; Milesi, C.; Hashimoto, H.; Wang, W.; Saatchi, S.; Yu, Y.; Myneni, R.B. Estimation of forest aboveground biomass in California using canopy height and leaf area index estimated from satellite data. *Remote Sens. Environ.* **2014**, *151*, 44–56.

20. Pflugmacher, D.; Cohen, W.B.; E. Kennedy, R. Using Landsat-derived disturbance history (1972–2010) to predict current forest structure. *Remote Sens. Environ.* **2012**, *122*, 146–165.

21. Zhang, X.; Kondragunta, S. Estimating forest biomass in the USA using generalized allometric models and MODIS land products. *Geophys. Res. Lett.* **2006**, *33*, L09402.

22. Jenkins, J.C.; Chojnacky, D.C.; Heath, L.S.; Birdsey, R.A. National-Scale biomass estimators for united states tree species. *For. Sci.* **2003**, *49*, 12–35.

23. Jenkins, J.C.; Chojnacky, D.C.; Heath, L.S.; Birdsey, R.A. *Comprehensive Database of Diameter-Based Biomass Regressions for North American Tree Species*; USDA Forest Service: Pennsylvania, USA, 2004; p. 48.

24. Jenkins, J.C.; Birdsey, R.A.; Pan, Y. Biomass and NPP estimation for the mid-atlantic region (USA) using plot-level forest inventory data. *Ecol. Appl.* **2001**, *11*, 1174–1193.

25. Ahmed, R.; Siqueira, P.; Hensley, S.; Bergen, K. Uncertainty of forest biomass estimates in North temperate forests due to allometry: Implications for remote sensing. *Remote Sens.* **2013**, *5*, 3007–3036.

26. Yoon, T.K.; Park, C.W.; Lee, S.J.; Ko, S.; Kim, K.N.; Son, Y.; Lee, K.H.; Oh, S.; Lee, W.K.; Son, Y. Allometric equations for estimating the aboveground volume of five common urban street tree species in Daegu, Korea. *Urban For. Urban Green.* **2013**, *12*, 344–349.

27. Paul, K.I.; Roxburgh, S.H.; Ritson, P.; Brooksbank, K.; England, J.R.; Larmour, J.S.; Raison, R.J.; Peck, A.; Wildy, D.T.; Sudmeyer, R.A.; *et al.* Testing allometric equations for prediction of above-ground biomass of mallee eucalypts in southern Australia. *For. Ecol. Manag.* **2013**, *310*, 1005–1015.

28. Zhao, F.; Guo, Q.; Kelly, M. Allometric equation choice impacts LiDAR-based forest biomass estimates: A case study from the Sierra National Forest, CA. *Agric. For. Meteorol.* **2012**, *165*, 64–72.

29. Ter-Mikaelian, M.T.; Korzukhin, M.D. Biomass equations for sixty-five North American tree species.*For. Ecol. Manag.* **1997**, *97*, 1–24.

30. Hyde, P.; Nelson, R.; Kimes, D.; Levine, E. Exploring LiDAR–RaDAR synergy—predicting aboveground biomass in a southwestern ponderosa pine forest using LiDAR, SAR and InSAR. *Remote Sens. Environ.* **2007**, *106*, 28–38.

31. Popescu, S.C. Estimating biomass of individual pine trees using airborne LiDAR. *Biomass Bioenergy***2007**, *31*, 646–655.

32. Sonter, L.J.; Moran, C.J.; Barrett, D.J.; Soares-Filho, B.S. Processes of land use change in mining regions. *J. Clean. Prod.* **2014,** *84*, 494–501.

33. Cheng, W.; Bian, Z.; Dong, J.; Lei, S. Soil properties in reclaimed farmland by filling subsidence basin due to underground coal mining with mineral wastes in China. *Trans. Nonferrous Met. Soc. China* **2014,***24*, 2627–2635.

34. Pietrzykowski, M.; Daniels, W.L. Estimation of carbon sequestration by pine (*Pinus sylvestris* L.) ecosystems developed on reforested post-mining sites in Poland on differing mine soil substrates.*Ecol. Eng.* **2014**, *73*, 209–218.

35. Strong, W.L. Vegetation development on reclaimed lands in the Coal Valley Mine of western Alberta, Canada. *Can. J. Bot.* **2000**, *78*, 110–118.

36. Russell, W.B.; La Roi, G.H.; Natural, R.O.I. Natural vegetation and ecology of abandoned coal-mined land, Rocky Mountain Foothills, Alberta, Canada. *Can. J. Bot.* **1986**, *64*, 1286–1298.

37. Casey, J. Climate, global warming, and daylight charts and data. *World Clim.* **2010,** *1*, 1–3.

38. Sanz-Cortiella, R.; Llorens-Calveras, J.; Escolà, A.; Arnó-Satorra, J.; Ribes-Dasi, M.; Masip-Vilalta, J.; Camp, F.; Gràcia-Aguilá, F.; Solanelles-Batlle, F.; Planas-DeMartí, S.; *et al.* Innovative LIDAR 3D dynamic measurement system to estimate fruit-tree leaf area. *Sensors* **2011**, *11*, 5769–5791.

39. Rosell, J.R.; Llorens, J.; Sanz, R.; Arnó, J.; Ribes-Dasi, M.; Masip, J.; Escolà, A.; Camp, F.; Solanelles, F.; Gràcia, F.; *et al.* Obtaining the three-dimensional structure of tree orchards from remote 2D terrestrial LiDAR scanning. *Agric. For. Meteorol.* **2009**, *149*, 1505–1515.

40. Nlungu-Kweta, P.; Leduc, A.; Bergeron, Y. Conifer recruitment in trembling aspen (Populus Tremuloides Michx.) stands along an east-west Gradient in the Boreal Mixedwoods of Canada. *Forests***2014**, *5*, 2905–2928.

41. Wylie, B.; Rigge, M.; Brisco, B.; Murnaghan, K.; Rover, J.; Long, J. Effects of disturbance and climate change on ecosystem performance in the Yukon River Basin Boreal Forest. *Remote Sens.* **2014**, 6, 9145–9169.

42. Carlson, T.N.; Riziley, D.A. On the relation between NDVI, fractional vegetation cover, and leaf area index. *Remote Sens. Environ.* **1997**, *62*, 241–252.

43. Turner, D.P.; Cohen, W.B.; Kennedy, R.E.; Fassnacht, K.S.; Briggs, J.M. Relationships between leaf area index and Landsat TM spectral vegetation indices across three temperate zone sites. *Remote Sens. Environ.* **1999**, *68*, 52–68.

44. Jiang, Z.; Huete, A.; Didan, K.; Miura, T. Development of a two-band enhanced vegetation index without a blue band. *Remote Sens. Environ.* **2008**, *112*, 3833–3845.

45. Huete, A.R. A Soil-Adjusted Vegetation Index (SAVI). *Remote Sens. Environ.* **1988**, *25*, 295–309.

46. Huete, A.; Didan, K.; Miura, T.; Rodriguez, E.P.; Gao, X.; Ferreira, L.G. Overview of the radiometric and biophysical performance of the MODIS vegetation indices. *Remote Sens. Environ.* **2002**, *83*, 195–213.

47. Schnur, M.T.; Xie, H.; Wang, X. Estimating root zone soil moisture at distant sites using MODIS NDVI and EVI in a semi-arid region of southwestern USA. *Ecol. Inform.* **2010**, *5*, 400–409.

48. Hunt, E.R., Jr.; Daughtry, C.S.T.; Eitel, J.U.H.; Long, D.S. Remote sensing leaf chlorophyll content using a visible band index. *Agron. J.* **2011**, *103*, 1090–1099.

49. Hunt, E.R.; Doraiswamy, P.C.; McMurtrey, J.E.; Daughtry, C.S.T.; Perry, E.M.; Akhmedov, B. A visible band index for remote sensing leaf chlorophyll content at the canopy scale. *Int. J. Appl. Earth Obs. Geoinf.* **2013**, *21*, 103–112.

50. Schowengerdt, R.A. *Remote Sensing: Models and Methods for Image Processing*, 3rd, ed.; Elsevier: Amsterdam, The Netherlands, 2007; p. 558.

51. Masoud, A.A.; Koike, K. Arid land salinization detected by remotely-sensed landcover changes: A casc study in the Siwa region, NW Egypt. *J. Arid Environ.* **2006**, *66*, 151–167.

52. Lambert, M.C.; Ung, C.H.; Raulier, F. Canadian national tree aboveground biomass equations. *Can. J. For. Res.* **2005**, *35*, 1996–2018.

53. Ung, C.H.; Bernier, P.; Guo, X.J. Canadian national biomass equations: new parameter estimates that include British Columbia data. *Can. J. For. Res.* **2008**, *38*, 1123–1132.

54. Zhang, J.; Huang, S.; Hogg, E.H.; Lieffers, V.; Qin, Y.; He, F. Estimating spatial variation in Alberta forest biomass from a combination of forest inventory and remote sensing data. *Biogeosciences* **2014**,*11*, 2793–2808.

55. Webster, R.; Oliver, M.A. *Geostatistics for Environmental Scientists*, 2nd, ed.; John Wiley & Sons, Ltd: Chichester, UK, 2007; p. 333.

56. Anderson, D.R. *Model Based Inference in the Life Sciences: A Primer on Evidence*; Springer Science + Business Media, LLC: New York, NY, USA, 2008; p. 203.

57. Akaike, H. A new look at the statistical model identification. *IEEE Trans. Autom. Control.* **1974**, *19*, 716–723.

58. Badreldin, N.; Goossens, R. Monitoring land use/land cover change using multi-temporal Landsat satellite images in an arid environment: A case study of El-Arish, Egypt. *Arab. J. Geosci.* **2013**, *7*, 1671–1681.

59. Kelarestaghi, A.; Jeloudar, Z.J. Land use/cover change and driving force analyses in parts of northern Iran using RS and GIS techniques. *Arab. J. Geosci.* **2009**, *4*, 401–411.

60. Schlagel, J.D.; Newton, C.M. A GIS-Based statistical method to analyze spatial change. *Photogramm. Eng. Remote Sens.* **1996**, *62*, 839–844.

61. Gilbert, R.O. *Statistical Methods for Environmental Pollution Monitoring*; Van Nostrand Reinhold: New York, NY, USA, 1987; p. 320.

62. Woodcock, C.E.; Macomber, S.A.; Kumar, L. Vegetation mapping and monitoring. In *Environmental Modelling with GIS and Remote Sensing*; Skidmore, A., Ed.; Taylor & Francis Group: London, UK, 2002; pp. 97–120.

# CHAPTER 11

# Fatigue Behavior of a Box-Type Welded Structure of Hydraulic Support Used in Coal Mine

**Xiaohui Zhao [1,2], Fuyong Li [2], Yu Liu [3,*] and Yanjun Fan [1]**

[1]*Key Laboratory of Automobile Materials, School of Materials Science and Engineering, Jilin University, Changchun 130025, China*

[2]*Zhengzhou Coal Mining Machinery Group Company Limited, Zhengzhou 450016, China*

[3]*School of Mechanical Science and Engineering, Jilin University, Changchun 130025, China*

## ABSTRACT

Hydraulic support is the main supporting equipment of the coal mining systems, and they are usually subjected to fatigue failure under the high dynamic load. The fracture positions are generally at welded joints where there is a serious stress concentration. In order to investigate and further improve the fatigue strength of hydraulic support, the present work first located the possible position where fatigue failure occurs through finite element analysis, and then fatigue tests were carried out on the different forms of welded joints for the dangerous parts. Finally, Fatigue strength-life (*S-N*) curves and fracture mechanism were studied. This research will provide a theoretical reference for the fatigue design of welded structures for hydraulic support.


## KEYWORDS

hydraulic support; finite element; stress concentration; weld; fatigue


## 1. INTRODUCTION

Coal represents the main energy source in the whole world. How to safely and efficiently exploit coal is a key issue [1]. As the main supporting equipment of the coal mining systems, hydraulic support is used to sustain the pressure of coal seams, which is the basic structure to ensure the safety of mining work. With the rapid development of mining technology as well as the large demand of raw coal, hydraulic support is developing towards the high-end and super high-end direction. Hydraulic support contains some box-type structures with a lot of stiffener plates. Welding is the main method to connect these plates into box-type structures. As is known, hydraulic support works under very large dynamic load, which might lead to the occurrence of fatigue failure. Moreover, fatigue failure is sensitive to structural stress concentration. Therefore, it will be worse if the structural stress concentration is located in welded joints. The reason is that welded joint is the weak section in the whole support due to the existence of welded defects. The fatigue performance of welded joint will significantly reduce under the action of structural stress concentration and welded defects.

At present, the service life of the domestic high-end hydraulic support is about 40,000 cycles under actual working load. According to the practical requirements, the service life of the domestic high-end hydraulic support should reach up to 80,000–100,000 cycles in the near future. Structural stress concentration and welded defects as the main factors to influence the fatigue life of hydraulic support, which should be paid high attention during the fracture and fatigue-related failure assessments of welded joints of hydraulic support [2,3]. Therefore, the fatigue performance of welded joints has to be considered when designing the fatigue strength of welded structures [4,5,6,7].

Stress concentration as the main factor affecting the fatigue performance of welded structure has to been considered. Huo and Masubuchi have been studying the assessment of fracture and fatigue-related failure [4,8]. Other scientists have also conducted a lot of research about the influence of stress concentration on the fatigue performance of welded structure. In addition to experimental studies, numerical simulations were also used to thoroughly analyze the stress concentration in welded structures [9,10]. Therefore, taking certain domestic hydraulic supports for example, this paper aimed at investigating and further improving the fatigue strength of hydraulic supports. The major effort first located the position that fatigue failure most likely to happen through finite element analysis, and then carried out fatigue tests for these dangerous parts, followed by studying the fatigue *S-N* curves and fracture mechanism. Taking the joints with low fatigue strength into consideration, the forms of welded joints were accordingly optimized. The current research as a combination of finite element simulation and experimental analysis will provide a theoretical reference for the fatigue design of hydraulic support.

## 2. MATERIAL AND EXPERIMENTAL PROCEDURES

### 2.1. Numerical Modeling

Hydraulic support in this study was shield-type, which consists of the base, connecting rods, shield beams, top beams, pillars, jacks and control facilities. Considering the complex structure of hydraulic supports, the geometrical model was built by three-dimensional modeling software, and then geometrical model was imported into the finite element analysis software for further computation. Base parts were set up first, which was regarded as the design basis. In order to successfully generate mesh and save computing time, the geometrical model was reasonably simplified on the premise that stress state of the main structures was not obviously affected. The principles were as follows [11]: (a) The location of the axes of each part remained unchanged, and the sizes of principal bearing parts were consistent with those in the original design; (b) the details of dangerous parts were retained; (c) the tiny process structures, such as small chamfers and holes that had little influence on stress state were omitted; (d) the equilibrium jack was simplified as a solid pillar, the length of which was determined by the working height of hydraulic support.

Hydraulic support usually bears the supporting load of pillar and the pressure load of working face with surrounding rock. Generally, the blocks are placed in different positions of the support in order to simulate different working conditions in the strength test.

In present finite element analysis, one of the most dangerous working condition is concerned, in which the load is combined by torsion load on the top beam and load on both ends of the base [12]. The specific loading form and position are shown in Figure 1.

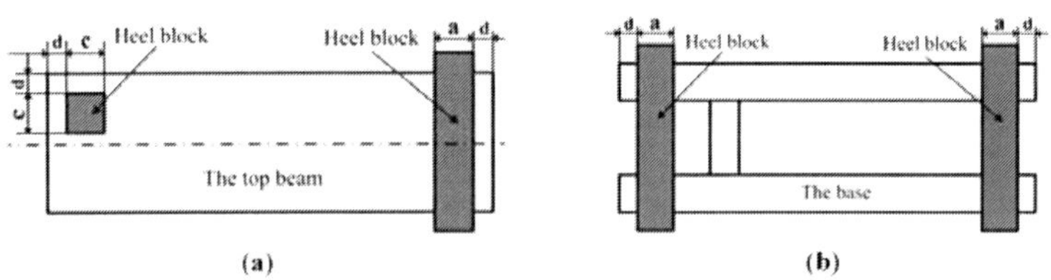


**Figure 1.** The load is combined by torsion load on the top beam and load on both ends of the base. (**a**) The size of heel block on the top beam ($a$ = 150 mm; $c$ = 300 mm; $d$ = 50 mm; thickness, $h$ = 50 mm); (**b**) The size of heel block under the beam ($a$ = 150 mm; $d$ = 50 mm; thickness, $h$ = 50 mm).

According to Chinese standards GB-25974.1-2010 [12], the load should be carried out on heel block. However, it is difficult to decide the surface load of heel block based on the nominal working resistance of pillars. Besides, the statically indeterminate problem may occur in finite element analysis. Thus, the pillars were divided into two parts close to the top and bottom nests. When the load is combined by torsion load on the top beam and load on both ends of the base, the height of hydraulic support is calculated according to the following formulas [12].

In this working condition, the minimum height ($H_{min}$) of hydraulic support is:

$$H_{min} = 3000\,\text{mm} \tag{1}$$

The maximum height ($H_{max}$) of hydraulic support is:

$$H_{max} = 4400\,\text{mm} \tag{2}$$

The distance ($L$) of hydraulic support is:

$$L = H_{max} - H_{min} = 4400\,\text{mm} - 3000\,\text{mm} = 1400\,\text{mm} \tag{3}$$

Therefore, the testing height ($H$) under this working condition is:

$$H = H_{max} - L/3 = 4400 - 1400/3\ \text{mm} = 3934\ \text{mm} \tag{4}$$

It has been known that if the load of heel block acting on hydraulic support was considered an external load, the heel block-support system will become a statically indeterminate system, which cannot be solved by equilibrium Equations. Therefore, the load of heel block acting on the support was considered as boundary conditions, while the nominal working resistance of pillars was imposed on the cross section of pillars near the base and top beam.

According to Chinese standards GB-25974.1-2010 [12], the rated working resistance ($F_r$) of single column for this type of hydraulic support is 6000 kN. The actual load ($F_a$) is 1.2 times of the rated load.

$$F_a = 1.2 \times F_r = 1.2 \times 6000\,\text{kN} = 7200\,\text{kN} \quad (5)$$

The cross-sectional area of live column and outer cylinder is:

$$S_1 = \frac{\pi d_1^2}{4} = \frac{\pi \cdot 180^2}{4}\text{mm}^2 \quad (6)$$

$$S_2 = \frac{\pi d_2^2}{4} = \frac{\pi \cdot 300^2}{4}\text{mm}^2 \quad (7)$$

The surface load of live column and outer cylinder is:

$$P_1 = \frac{F' \times 10^3}{S_1} = \frac{7.2 \times 10^6}{\frac{\pi \cdot 180^2}{4}}\text{mm}^2 = 283\text{ MPa} \quad (8)$$

$$P_2 = \frac{F' \times 10^3}{S_2} = \frac{7.2 \times 10^6}{\frac{\pi \cdot 300^2}{4}}\text{mm}^2 = 102\text{ MPa} \quad (9)$$

Therefore, the actual working resistance of single pillar is $F_a = 7200$ kN. Then, the surface load of live column and outer cylinder rafter converted is 283 and 102 MPa, respectively. The load and boundary conditions of hydraulic support are shown in Figure 2. In this study, the main material of hydraulic support is Q690 steel. In the process of analysis, the double linear constitutive model was adopted for plastic analysis (see Figure 3). The entire geometrical model was meshed with eight-node linear brick elements Soild185.

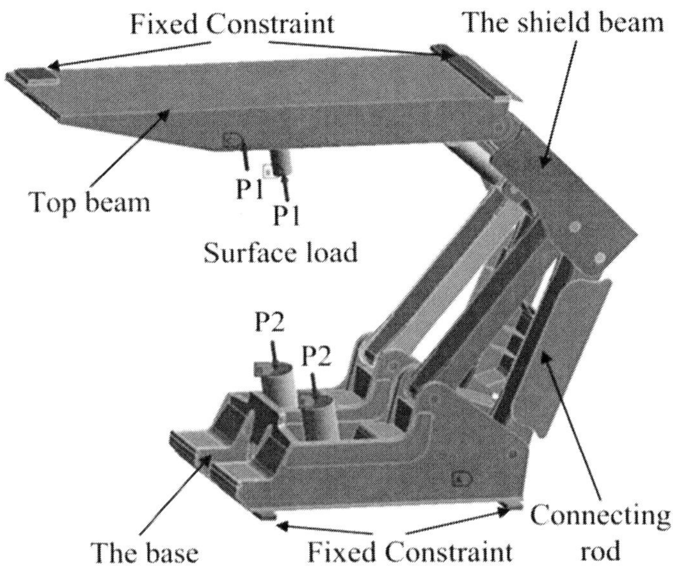


**Figure 2.** The load and boundary conditions of hydraulic support.

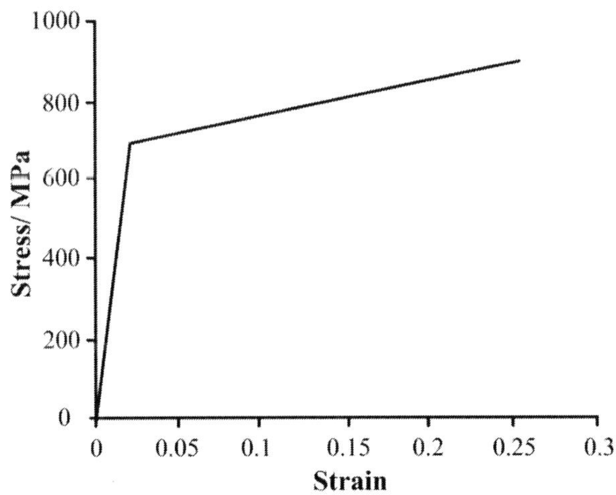


**Figure 3.** The double linear curve of Q690 steel.

## 2.2. Fatigue Testing

T-joints are widely used in the connections among main parts (main reinforcement and roof, main reinforcement and cover plate, transverse reinforcement plate and main reinforcement). For the T-joints, fatigue tests are hard to be carried out. Thus, the cruciform welded joint was used to simulate T-joint to carry out fatigue tests. One side of the cruciform welded joint conforms to the characteristics of the T-joint, and the sample can be clamped conveniently in the process of fatigue tests.

Fatigue tests were conducted on 200 kN high frequency fatigue testing machine (CIMACH, Changchun, China) with static load error for full measuring range of ±0.2% and dynamic load error of ±2%. Under tension–tension constant amplitude loading mode, each type of welded joint was tested at five stress levels with stress ratio $R = 0.1$ at room temperature in air environment.

## 2.3. The Design of Welded Joint

Fatigue performance of welded structure mainly depends on the dimensional details of welded joint. In order to improve the fatigue performance of hydraulic support, a better form of welded joint must be designed. The main body of hydraulic support involves welded joints (main reinforcement and roof, main reinforcement and flats, transverse reinforcement plate and main muscle). These welded joints are basic T-type joints. The forms of T-joints can be bilateral symmetry or unilateral symmetry. While designing the welded joint, these two conditions were both covered.

For the bilateral symmetry welded joint: one was cruciform joint with fillet weld (weld leg size of 14, 16 and 18 mm, respectively); and another was cruciform joint with full penetration weld (weld leg size of 8 mm), which had the nearly same effective sectional area with that of weld leg size of 18 mm (Figure 4). The unilateral symmetry welded joints with different size are schematically shown in Figure 5.

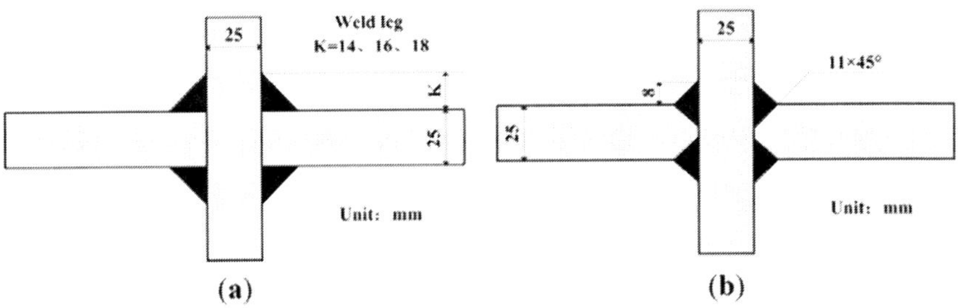


**Figure 4.** The size and details of bilateral symmetry welded joint. (**a**) Welded joint without groove in $K = 14$, 16 and 18 mm; (**b**) Welded joint of 11 × 45° groove and 8 mm weld leg.

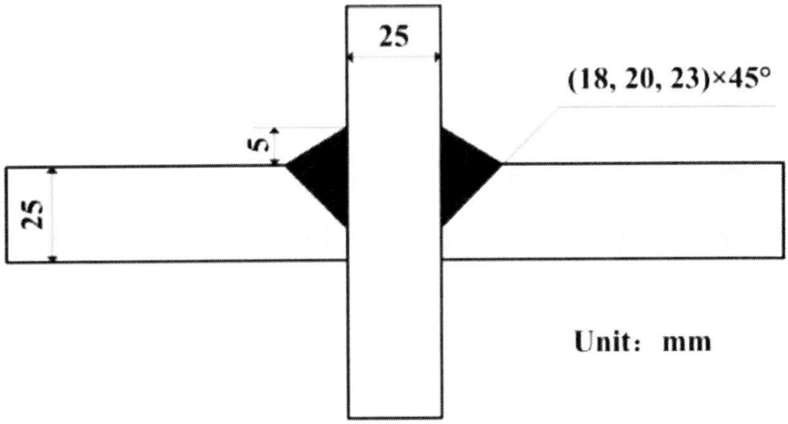


**Figure 5.** The size and details of unilateral symmetry welded joints.

# 3. RESULTS AND DISCUSSIONS

## 3.1. The Stress Field of Finite Element Simulation

The top beam of hydraulic support is welded structure with box shape, which plays the action to support the working surface of the roof in order to prevent gangue and roof coal caving. Figure 6 shows the stress contour of hydraulic support under the condition of the top beam subjected to torsion load and both ends of the base subjected to working load. Stress distribution is very uneven in Figure 6, but most parts present symmetrical distribution. Obviously, the higher stress appears in the top beam and the base. The maximum stress is 867.15 MPa in the top beam.

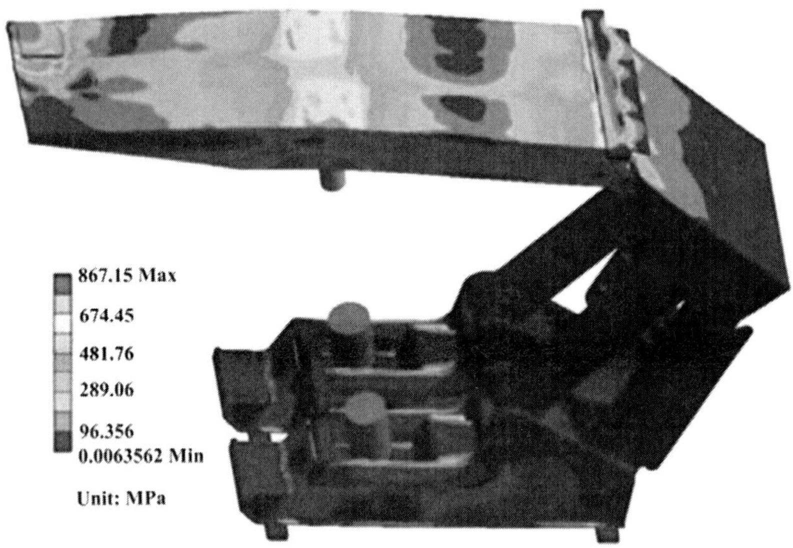


**Figure 6.** The stress contour of the entire hydraulic support.

The detailed stress distribution of the base is shown in Figure 7.

The stress distribution in both sides of the base is symmetrical with the maximum value of 758.52 MPa. Stress concentration occurs in the border area between bottom plate and main reinforcement or coverplate and main reinforcement. Meanwhile, the stress in the contact zone between the column nest piece and right stiffener plate is also larger.

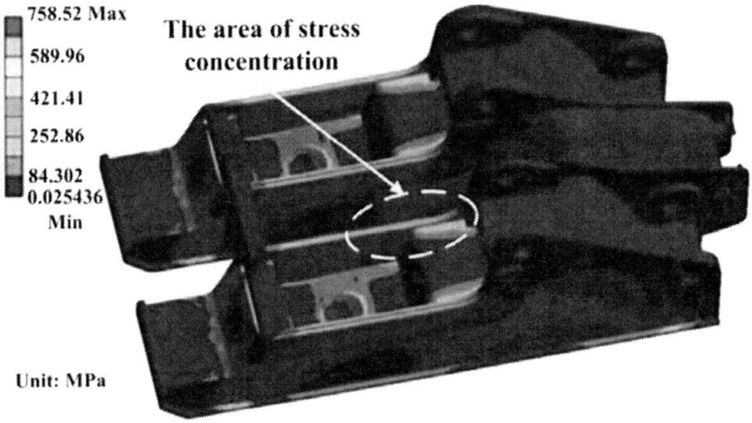


**Figure 7.** The stress distribution in the base.

The top beam suffers more severe load under this working condition, as partial load effect is also involved in torsion process. Figure 8 and Figure 9 show the stress distribution of the front and back sides of the top beam, respectively. It can be seen in Figure 8 and Figure 9 that the area of high stress is larger in this working condition. The maximum stress of 812.24 MPa appears near the block. The main stress concentration area mainly distributes in both sides and middle area of the top beam.

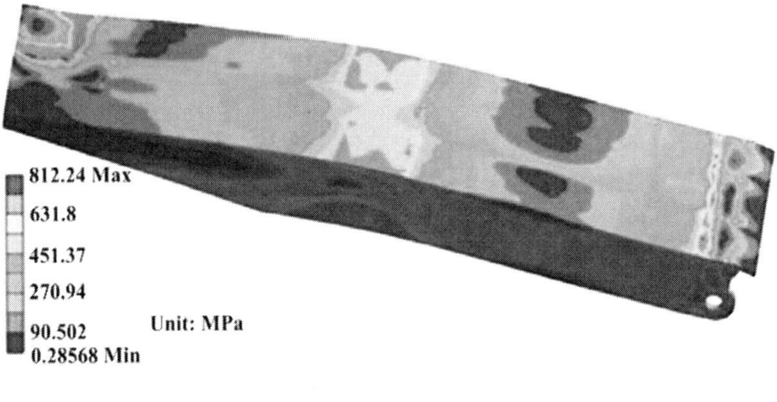


**Figure 8.** The stress distribution of the front side of the top beam.

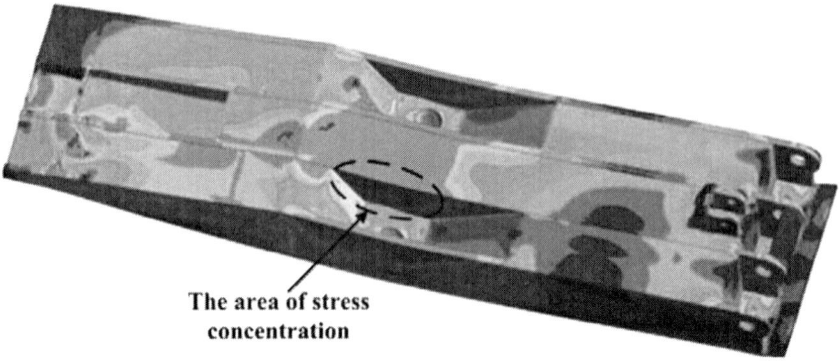


**Figure 9.** The stress distribution of column nests of the back side of the top beam.

Figure 10 indicates that the high stress distribution of shield beam. The maximum stress of 790.25 MPa exists in the pin hole connecting to the jack. Meanwhile, the area between coverplate and main stiffened plate also exists stress concentration.

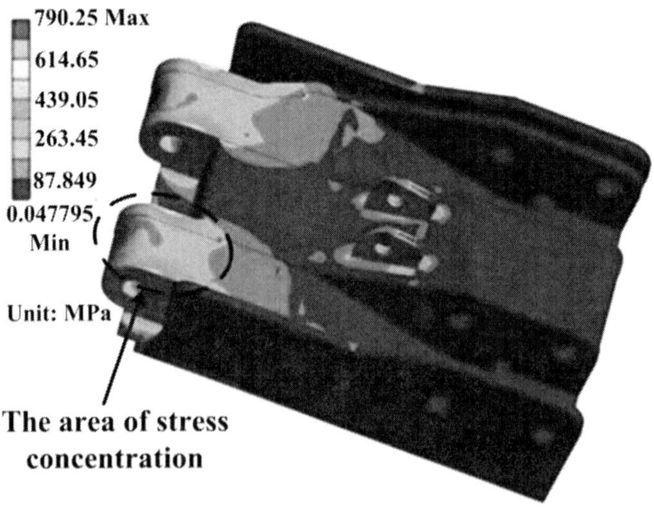


**Figure 10.** The stress distribution of shield beam.

Simulation results show that the severe stress concentration universally appears in the welded position of the main body. From Figure 7, Figure 9 and Figure 10, it can be seen that the area of stress concentration mainly exists in the base, top beam and shield beam. It is known that the compressive stress can inhibit crack formation and propagation, while the tensile stress is beneficial to crack formation and propagation. For a large welded structure, under the action of external loading, some weld seams bear tensile load, other weld seams bear compressive load. Generally, the weld seam under compressive load will be not easy to crack during service. However, the bearing capacity of weld seam under tensile external load and welded residual tensile stress will be reduced. Figure 11 shows the actual damaged conditions of the base, top beam and shield beam of hydraulic support during actual service. It is observed that the fracture locates at weld toe of the box-shape welded structure, which is consistent with the simulation results. The areas of stress concentration (the actual fracture positions) are indicated by arrow marks in Figure 7, Figure 9 and Figure 10. Therefore, it is important to choose an appropriate form of welded joint to improve the fatigue performance of hydraulic support.

**Figure 11.** The damaged positions of hydraulic support during actual service.

## 3.2. Fatigue S-N Curve

The stress level and fatigue life together determine the fatigue performance of the sample. The relationship between applied stress and fatigue life is represented in Equation (10). The curve based on Equation (10) is the traditional *S-N* curve.

$$N = \frac{C}{\Delta\sigma^{m}} \tag{10}$$

According to *S-N* curve, the fatigue strength corresponding to a cyclic number can be obtained. Fatigue testing data should be analyzed according to the principle of statistical method established in IIW (International Institute of Welding) [13] by Hobbacher. The authorized surviving fraction is 50%, while the confidence is 75%.

Nominal value is calculated as the following process:

Calculating all the stress range $\Delta\sigma$ and the cycle index $N$ (in logarithm with the base-number 10) of fatigue testing data points;

Calculating the exponent $m$ and the constant $\log C$ regressively, using the power function model:

$$m \log \Delta\sigma + \log N = \log C \tag{11}$$

Fatigue *S-N* curves of the bilateral symmetry welded joints are shown in Figure 12.

By analyzing *S-N* curves in Figure 12, the fatigue performance of joints with different types can be compared. Table 1 presents the maximum stress corresponding to cycle numbers of 30,000, 60,000 and 90,000, respectively. The numbers given in Table 1 are taken from the *S-N* curves in Figure 12 obtained by regression analysis.

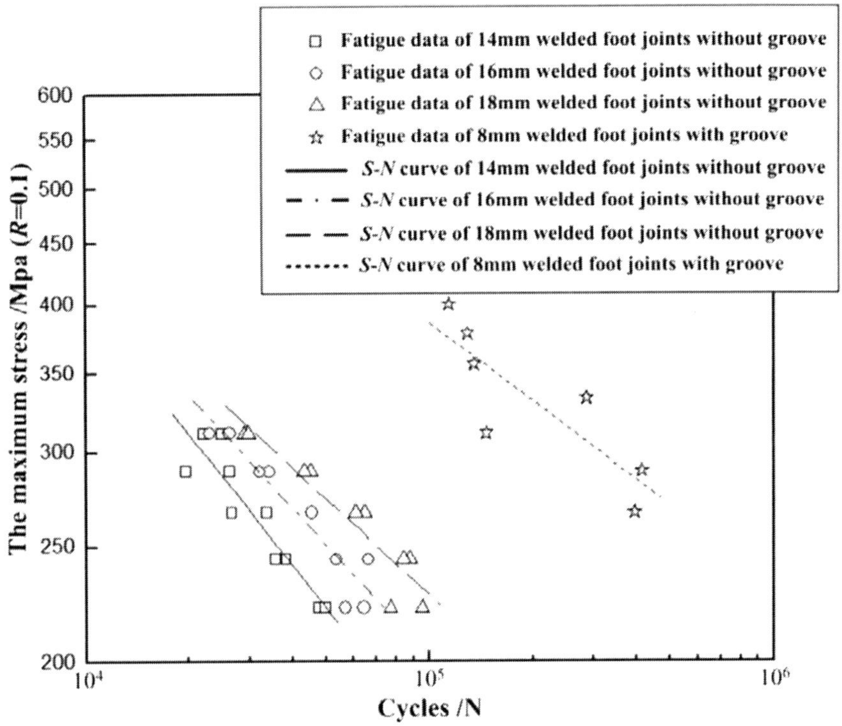


**Figure 12.** Fatigue *S-N* curves of the bilateral symmetry welded joints.

**Table 1.** Maximum stress corresponding to different cycle numbers.

| Cycle Times | Without Groove + 14 mm Weld Leg | Without Groove + 16 mm Weld Leg | Without Groove + 18 mm Weld Leg | Without Groove + 8 mm Weld Leg |
|---|---|---|---|---|
| 30,000 | 268 MPa | 295 MPa | 315 MPa | 501 MPa |
| 60,000 | 206 MPa | 238 MPa | 262 MPa | 429 MPa |
| 90,000 | 179 MPa | 208 MPa | 234 MPa | 393 MPa |

It can be observed that fatigue life of the fillet welded joint with bilateral symmetry groove increases with the increment of weld leg size. Compared with that of 16 mm fillet welded joint, fatigue strength of the 18 mm fillet welded joint has been enhanced by 6%, 10% and 12.5% corresponding to cycle numbers of 30,000, 60,000 and 90,000, respectively. For the fully penetrated fillet welded joint with groove of

11 × 45° and weld leg size of 8 mm, compared with that of 18 mm fillet welded joint, fatigue strength has been significantly improved by 59%, 63% and 68% corresponding to cycle number of 30,000, 60,000 and 90,000, respectively. Clearly, the fully penetrated fillet welded joint with groove is extremely advantageous.

*S-N* curves of the unilateral symmetry welded joints with the different size details are shown in Figure 13.

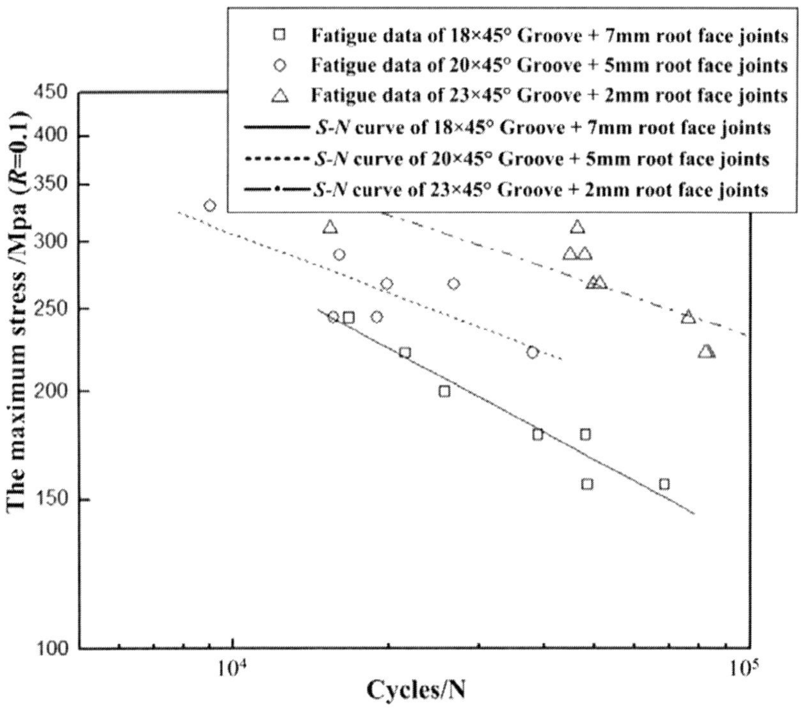


**Figure 13.** Fatigue *S-N* curves of the unilateral symmetry welded joints.

By analyzing the *S-N* curves in Figure 13, fatigue performance of the unilateral symmetry welded joints with different size details can be obtained. Table 2 shows the maximum stress corresponding to cycle numbers of 30,000, 60,000 and 90,000, respectively. The numbers given in Table 2 are also taken from the *S-N*curves in Figure 13 obtained by regression analysis.

It can be seen in Figure 13 and Table 2 that fatigue life of the unilateral symmetry welded joint increases as the size of root face decreases. Fatigue strength of the welded joint with groove size of 20 × 45° has been enhanced by 20.4%, 30.4% and 32.4% compared with that with groove size of 18 × 45° corresponding to cycle numbers of 30,000, 60,000 and 90,000, respectively. As for the fully penetrated welded joint with groove size of 23 × 45°, fatigue strength has been improved by 26.3%, 24.8% and 28.3% compared with that with groove size of 20 × 45° corresponding to cycle numbers of 30,000, 60,000 and 90,000, respectively. Therefore, in order to greatly improve fatigue life of the unilateral symmetry welded joint, grooving and full penetration is recommended.

**Table 2.** Maximum stress corresponding to different cycle numbers.

| Cycle Number | 18 × 45° Groove + 7 mm Root Face Non-Penetration | 20 × 45° Groove + 5 mm Root Face Non-Penetration | 23 × 45° Groove + 2 mm Root Face Full Penetration |
|---|---|---|---|
| 30,000 | 196 MPa | 236 MPa | 298 MPa |
| 60,000 | 158 MPa | 206 MPa | 257 MPa |
| 90,000 | 139 MPa | 184 MPa | 236 MPa |

## 3.3. Fatigue Fracture Location

### 3.3.1. The Bilateral Symmetry Welded Joints

Figure 14a shows that fatigue fracture is absolutely from the position where weld root was incompletely penetrated.

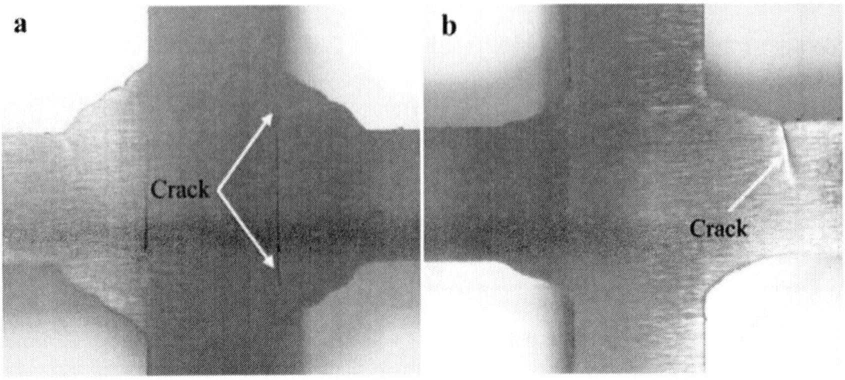


**Figure 14.** The position of crack initiation for the bilateral symmetry welded joints. (**a**) fatigue fracture is from weld root; (**b**) fatigue fracture is from weld toe.

For the fully penetrated welded joint with groove, fatigue fracture is caused by the stress concentration at weld toe (see Figure 14b). Obviously, by grooving and fully penetrating, fatigue crack initiations all appear at weld toe where stress concentration is relatively serious. Although some defects still exist at weld root (caused by manipulation or groove size), which will not produce fatigue crack. For the bilateral symmetry welded joints without groove, crack initiation usually develops at weld root of the fillet welded joint. However, for the fully penetrated welded joint with groove, fatigue crack initiations all form at weld toe. Fatigue performance cannot be improved through post-processing the weld toe if crack initiates at weld root. Yet, fatigue performance can be enhanced by reducing the stress concentration through grinding or remelting after welding if crack initiates at weld toe. Thus, the fully penetrated joint with groove is recommended for bilateral symmetry weld.

### 3.3.2. The Unilateral Symmetry Welded Joints

Fatigue cracks of the incompletely penetrated joints with groove size of 18 × 45° and root face of 7 mm as well as the incompletely penetrated joints with groove size of 20 × 45° and root face of 5 mm are all caused by the natural stress concentration at root face (see Figure 15a).

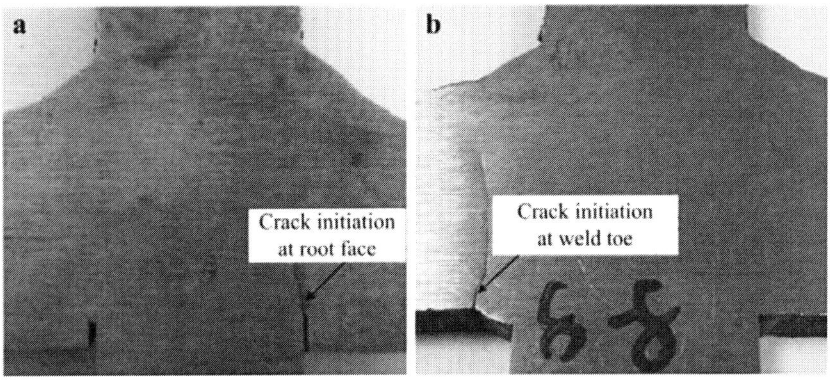


**Figure 15.** The position of crack initiation for the unilateral symmetry welded joints. (**a**) crack initiation is from root face; (**b**) crack initiation is from weld toe.

For the fully penetrated joints with groove size of 23 × 45° and root face of 2 mm, the root face of 2 mm has not existed after welding. Thus, the position of fatigue fracture will change from weld face into weld toe owing to the fully penetrated root face (see Figure 15b).

## 3.4. The Analysis of Conventional Fatigue Fracture Mechanism

For the bilateral symmetry welded joints, fatigue initiations of the joints without groove and the joints with groove are at weld root (Figure 16a) and weld toe (Figure 16b), respectively. The specific characteristics of crack initiation are shown in Figure 16.

Figure 17a indicates that the obvious fatigue striations have been generated for the defects caused by weld root. The reason is that the force along the direction of crack propagation at weld root is relatively small compared to that at weld toe. On the contrary, due to the larger stress at weld toe, crack extension rate under this state is relatively fast.

Thus, fatigue striations are not obvious. Only the trace of crack propagation can be observed in the low-magnification Microscope (Carl Zeiss Jena, Oberkochen, Germany) (see Figure 17b).

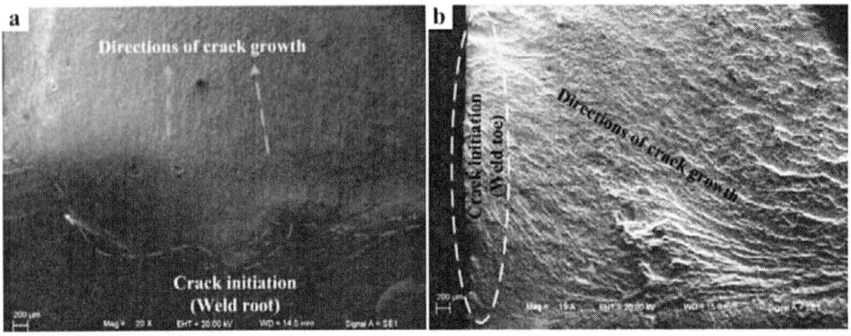


**Figure 16.** Crack initiation for the bilateral symmetry welded joints. (**a**) crack initiation is from weld root; (**b**) crack initiation is from weld toe.

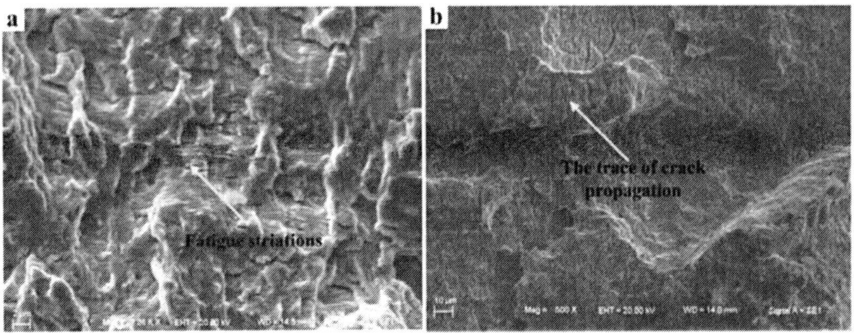


**Figure 17.** Fatigue striations in crack propagation region of the bilateral symmetry welded joints. (**a**) crack propagation region corresponding toFigure 16a; (**b**) crack propagation region corresponding to Figure 16b.

For the unilateral symmetry welded joints, with the decrease of the size of root face, penetration performance and fatigue resistance of joints have been significantly improved. The position of fatigue initiation also shifts from weld root to weld toe (see Figure 15b). Therefore, fatigue fracture position of the penetrated joint is at weld toe, and its fatigue crack initiation morphology is consistent with that in Figure 16b. Fatigue crack initiation region and fatigue crack propagation region of the incompletely penetrated joints are shown in Figure 18.

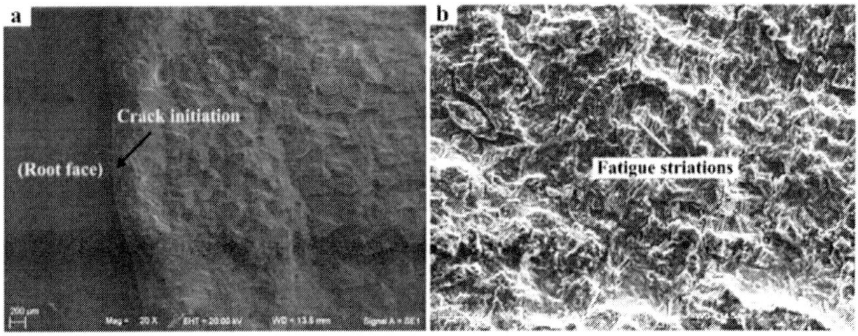


**Figure 18.** Fatigue crack initiation and fatigue striations of the incompletely penetrated joints for the unilateral symmetry welded joints. (**a**) crack initiation; (**b**) fatigue striation.

Fracture section of the instantaneous destruction region has the typical characteristic of dimple, which indicates that the fracture belongs to ductile fracture. In particular, the second phase particles still exist in the dimples, showing excellent ductility characteristics (see Figure 19).

# 4. CONCLUSIONS

By analyzing the structural characteristics of hydraulic supports based on simulation and fatigue testing, we can draw the following conclusions:

(1) Simulation results are consistent with the actual damaged positions of hydraulic supports.

(2) Structural stress concentration is the main factor leading to the fatigue damage of hydraulic supports. The key point of fatigue design of welded structure is to avoid or reduce stress concentration.

(3) This research provides a theoretical reference and a certain guiding significance for the fatigue design of hydraulic supports.

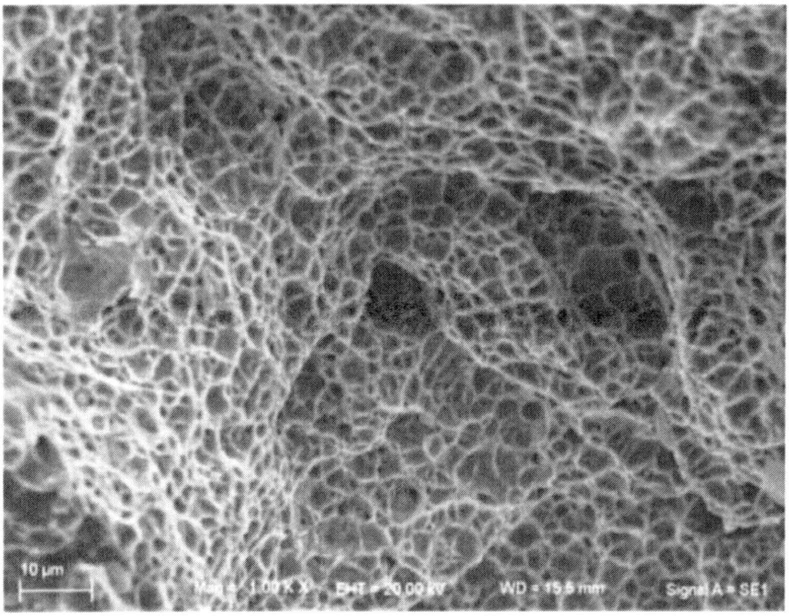


**Figure 19.** Fracture section of the instantaneous destruction region.

## ACKNOWLEDGMENTS

The authors acknowledge financial support by the National Science Foundation through Grant No. 51305160 and 51405182, the Research Funds for the Central Universities (JCKY-QKJC17), and Project 2015071 Supported by Graduate Innovation Fund of Jilin University.

## AUTHOR CONTRIBUTIONS

Yu Liu conceived and designed the experiments; Xiaohui Zhao, Fuyong Li and Yanjun Fan performed the experiments, analyzed the data and wrote the paper.

# REFERENCES


1. Wang, J.J.; Jia, X.L.; Fan, X. The strength finite element analysis of whole caving hydraulic support.*Coal Mine Mach.* **2005**, *3*, 36–38.

2. Zheng, X.W.; Zhang, H.; Liu, J.L.; An, X.L. Fatigue life analysis of hydraulic support based on FEM.*Min. Mach.* **2010**, *38*, 7–10.

3. Wu, L.B. The development and outlook of hydraulic support. *J. Huinan Vocat. Tech. Coll.* **2006**, *6*, 44–45.

4. Huo, L.X. *The Fracture Behavior and Evaluation of the Welding Struct*; Machinery Industry Press: Beijing, China, 2000.

5. Kirkhope, K.J.; Bell, R.; Caron, L.; Basu, R.I.; Ma, K.-T. Weld detail fatigue life improvement techniques. Part 1: Review. *Mar. Struct.* **1999**, *12*, 447–474.

6. Habibi, N.; H-Gangaraj, S.M.; Farrahi, G.H.; Majzoobi, G.H.; Mahmoudi, A.H.; Daghigh, M.; Yari, A.; Moridi, A. The effect of shot peening on fatigue life of welded tubular joint in offshore structure.*Mater. Des.* **2012**, *36*, 250–257.

7. Tian, X.T. *The Welding Structure*; Machinery Industry Press: Beijing, China, 1982.

8. Masubuchi, K. *Analysis of Welded Structures*; Pergamon Press: Oxford, UK, 1980.

9. Radaj, D.; Sonsino, C.M.; Fricke, W. Recent developments in local concepts of fatigue assessment of welded joints. *Int. J. Fatigue* **2009**, *31*, 2–11.

10. Radaj, D. Review of fatigue strength assessment of nonwelded and welded structures based on local parameters. *Int. J. Fatigue* **1996**, *18*, 153–170.

11. Li, C.J. Computer Aided Engineering Analysis of Hydraulic Support. Master's Thesis, Shandong University, Jinan, China, 2005.

12. China Coal Industry Association. *Powered Support for Coal Mine-Part 1: General Specification*; Standardization Administration of the People's Republic of China: Beijing, China, 2011.

13. Hobbacher, A.F. The new IIW recommendations of fatigue assessment of welded joints and component—A comprehensive code recently updated. *Int. J. Fatigue* **2009**, *31*, 50–58.

# Index

## S



## T



## U



## V



## W